Making Connections

Communication through the Ages

Charles T. Meadow

The Scarecrow Press, Inc.
Lanham, Maryland, and London
2002

SCARECROW PRESS, INC.

Published in the United States of America
by Scarecrow Press, Inc.
4720 Boston Way, Lanham, Maryland 20706
www.scarecrowpress.com

4 Pleydell Gardens, Folkestone
Kent CT20 2DN, England

British Library Cataloguing-in-Publication Information Available

Library of Congress Cataloging-in-Publication Data

Meadow, Charles T.
   Making connections : communication through the ages / Charles T. Meadow.
     p. cm.
   Includes bibliographical references and index.
   ISBN 0-8108-4233-5 (alk. paper) — ISBN 0-8108-4234-3 (pbk. : alk. paper)
     1. Communication—History. I. Title.
   P90 .M363 2002
   302.2'09—dc21   2001054982

♾™ The paper used in this publication meets the minimum requirements of
American National Standard for Information Sciences—Permanence of
Paper for Printed Library Materials, ANSI/NISO Z39.48-1992.
Manufactured in the United States of America.

To the University of Toronto Libraries and the librarians
for their magnificent collections and wonderful service

Contents

List of Illustrations   ix
Preface   xiii

Part 1: The Basics of Communication   1
  1 Background   3
    Introduction; The Basic Questions; What Is Commu-
    nication?; What Is Information?; What Is a Medium?;
    How Is Information Represented?; Was McLuhan
    Right—Is the Medium the Message?

Part 2: Telecommunication before Steam and Electricity 31
  2 Spoken Language and Sound Transmission   33
    Introduction; Spoken Language; Sound-Based
    Technology
  3 Writing and Printing   47
    Introduction; Pictures and Pictorial Writing; Alpha-
    bet; Other Languages; Recording Media; The Tools
    of Writing; Photography; The Effect of Writing on
    Civilization
  4 Visual Signaling   77
    Introduction; Fire and Smoke; Flags, Semaphores,
    and Body Movements
  5 Transportation as Communication—1   95
    Introduction; Animals and Roads; Muscle-Powered
    and Sailing Ships

Contents

Part 3: Steam, Internal Combustion, and Electricity   107

  6 Transportation as Communication—2   109
    Introduction; Steam Power; Wingless Aircraft; Internal Combustion
  7 Telegraph   125
    Introduction; Beginnings; Telegraphs; Practical and Impractical; How It Works; Telegraphy Becomes a Business; Crossing the Oceans; The Industry Goes Downhill
  8 Telephone   145
    Introduction; In the Beginning Was the Word; Sound and Electricity; How the Telephone Works; Telephone as a Business; New Developments

Part 4: Electronics   179

  9 Radio   183
    Introduction; Beginnings; More Physics; Wireless Telegraphy; Radar; Radio Broadcasting; Social Effects of Radio Broadcasting; Newer Technologies
  10 Television   221
    Introduction; Beginnings; Technicalities; The Developers; Broadcasting; New Developments in Television; Impact
  11 Communication Satellites   263
    Introduction; The Nature of the Bird; History; Technicalities; Impact
  12 The Internet and the Information Highway   277
    Introduction; Technicalities; Cost; History; Impact

Part 5: Looking Backward and Forward   303

  13 Summation and Projection   305
    Introduction; Technologies that Accompanied Major Change; The Coming Changes; A Review of Principles That Do Not Change

14 One Hundred Dates to Remember   325

Bibliography   337
Index   353
About the Author   365

Illustrations

Figures

1.1	The basic elements of a communication system	6
1.2	Symbols used in print communication	7
1.3	A graphic road safety sign	10
1.4	An early cave drawing	17
1.5	The beginnings of airmail	19
1.6	The evolution of some letters	21
1.7	The name on the nose	22
1.8	A watch face combining analog and digital representations	24
1.9	Conveying the message with impact	28
1.10	Communication in the street	28
2.1	Symbols from American Sign Language	34
2.2	Duck talk	36
2.3	The head of a steel drum	40
2.4	An alpenhorn	42
2.5	Bagpipes on the battlefield	43
3.1	*Nude Descending a Staircase*	48
3.2	Evolution of cuneiform	51
3.3	An Iroquois wampum belt	52
3.4	Mathematical symbols	57
3.5	Musical symbols	57
3.6	Chemical symbols	58
3.7	Form of a Sumerian cuneiform stylus	64
3.8	A typewriter of the early twentieth century	66
3.9	The Linotype machine and its product	67

3.10	The camera obscura	70
4.1	The Washington family coat of arms	84
4.2	Military unit symbols	85
4.3	Some national flags	86
4.4	Semaphore flags	88
4.5	A railroad semaphore	89
4.6	The Chappé semaphore	90
4.7	Modern computer icons	92
5.1	A *travois*, freight and passenger carrier for unpaved roads and unwheeled vehicles	97
5.2	The Pony Express	101
5.3	Motive power in a Roman trireme, first century CE	102
5.4	The *Santa Maria*, fifteenth century	103
5.5	Sextant	103
5.6	A canal boat, nineteenth century	105
6.1	Hero's "steam engine "	110
6.2	The first successful steamship, 1807	112
6.3	The S.S. *United States*, undefeated transatlantic speed champion	112
6.4	Constructing the railroad and its helper, the telegraph	114
6.5	The *Graf Zeppelin*	117
6.6	An early Daimler motorcar	119
6.7	The Wright brothers' first airplane	121
6.8	The Douglas DC-3	122
7.1	The Cooke-Wheatstone telegraph	130
7.2	Morse code	132
7.3	Morse telegraph components	133
7.4	The Baudot code	135
8.1	Making waves in the water	151
8.2	Basic wave structure	152
8.3	A pure wave form	152
8.4	A voice spectrograph	153
8.5	A primitive sound-powered telephone	154
8.6	The first patented Bell telephone	155
8.7	Schematic diagram of a telephone transmitter	156
8.8	Schematic diagram of a telephone receiver	157
8.9	A nineteenth-century telephone "on hook "	159

8.10	Telephone lines abound in snowbound New York City	160
8.11	An early switchboard in 1879	162
8.12	Telephone networks	164
8.13	The number of telephones or lines in the United States, 1876-1996	167
8.14	Cellular telephone appears on the scene, literally	171
8.15	A wireless telephone	171
8.16	Routing of a cell-phone call	172
9.1	The electromagnetic spectrum	187
9.2	Line of sight transmission	189
9.3	Extending the range of waves by bouncing off the ionosphere	190
9.4	Magnetic induction	191
9.5	Marconi's coherer, an early amplifier	192
9.6	Basic concept of radar	200
9.7	Early form of radar antennas	201
9.8	A modern radar antenna on an old-fashioned ship	202
9.9	More modern radar antennas	203
9.10	The Edison Effect and the de Forest Audion	205
9.11	The number of radio broadcasting stations in the United States, 1921-1996	209
9.12	The new downtown skyline	215
10.1	The scanning of a graphic image	222
10.2	Transmitting the graphic image	223
10.3	Variation in the number of shades of gray in an image	224
10.4	Enlarged pixels	225
10.5	The Nipkow disk	229
10.6	The cathode-ray tube	232
10.7	A 30-line image	233
10.8	The number of commercial television broadcasting stations and cable systems in the United States, 1941-1996	242
10.9	The control room in a modern television studio	244
10.10	A mobile television unit	245
10.11	Separation of the color elements of a television image	247
10.12	The new shape of television?	253

10.13 Virtual television images   257
11.1   Coverage of the globe by communication satellites   264
11.2   A satellite in geosynchronous orbit   265
11.3   Earth station to earth station linkage   265
11.4   A modern communications satellite   267
11.5   A modern antenna for communication with a satellite 267
12.1   Packet switching   281
12.2   A tabulating-machine card   284
12.3   A console used in SAGE   285
12.4   The growth of the Internet   289
12.5   The growth of electronic mail   291

Table

6.1   Transatlantic crossing speeds   113

Preface

Our modern systems of telecommunication collectively represent the most important technological developments of the twentieth century in terms of their effect on society. Steam and electricity were harnessed in the nineteenth century. Atomic energy never really delivered on its promises of benefits to humankind. Computers are of great importance, of course, but much of their prominence came only after they became interconnected by communications lines. It is telecommunication that gives us the telephone, radio, television, automatic-teller machines, electronic mail, and the World Wide Web.

Taken literally, telecommunication means far or distant communication, although today it tends to mean, as the *Encyclopaedia Britannica* puts it, "transmission of an information-bearing electronic signal." A dictionary on my desk, published in 1960, does not even define the term, which suggests its lack of recognized importance as recently as that time. But people were inventing more effective ways to communicate, over distance and over time, ever since they began to communicate by gesture, sound, drawing, or playing music.

A great deal of ingenuity was shown by our ancestors who developed many of the modern principles of communication, if not the mechanics of *tele*communication, many centuries ago. Is the medium the message? That is an important concept, but can't be taken entirely literally. Messages much like those we send today were being sent millennia ago, before our present media of transmission were invented. The principles of how to *communicate*, if not necessarily how to *transmit*, evolved many years ago but there can be no doubt that the depiction of a war or athletic contest has far more impact if it comes via live television than via a poetic description long after the fact.

This book is primarily intended for a lay audience, that is, for readers who have little or no technical background in communications, but recognize its importance in our lives, the importance of the human beings who communicate, and the content of what they have to say, rather than merely the mechanical means of doing so. Some technical people may benefit from learning some of their field's history, as well.

Why should people be interested in a history of communication? Because it continues to affect us all, because there is a great deal of hype and misunderstanding about the modern systems. Modern communications technology is not all good in the social sense. We have the ability to keep up with business or the family from anywhere on earth, to have a medical condition diagnosed from afar, but also to fake television images and to distribute hate and pornography. Knowing how we got to where we are may help in understanding what is coming, what effect it may have, and how we want our systems to be used. And it is often a fascinating story. We will also see that communication is, and was, universal. Ideas came from all over the world—Asia, Africa, Europe, and America. Generally I have taken an American slant, for example by presenting U.S. statistics where such are used at all, or illustrating with American anecdotes. All this because I assume the readership will be primarily American and not to deny the enormous contributions made in other lands.

I am not a professional historian. I learned several lessons about history while doing the research for this book. First, we often do not know how some event in history actually took place, or exactly who was involved, and we may never find out. Who invented speech? What person invented the idea of a phonetic alphabet? Probably no one person in either case. We're not even certain which people invented the alphabet. Who actually first invented the telephone? Whose telephonic invention got to the U.S. Patent Office first? Those two questions may have different answers. Even the invention of wireless telegraphy, early radio, is disputed and the history of television is rife with law suits contesting the priority of invention.

My second lesson was that many inventions that we attribute to a single person were the result of group efforts or resulted from the work of a series of people, each contributing a bit more, until a single, now well-known name came into the picture. Samuel F. B. Morse did not invent the electric telegraph, nor did he develop his own telegraph entirely alone. But those facts need not detract from his enormous accomplishments.

The book starts with some fundamentals of communication, essential to understanding the nature and importance of some of the technologies involved. Then it proceeds in a combination of chronological order while also grouping similar inventions: starting with the earliest forms of writing, the alphabet, and printing. After movable-type printing, the developments in communication came mostly in the form of improved transportation so there is a review of transportation history where it affected communication. Early in the nineteenth century came steam and soon thereafter electricity, which revolutionized transportation and ushered in the age of high-speed telecommunication. The twentieth century brought us electronics and with it radio, television, communication satellites, and the computers and the Internet. We conclude with a summary and a hesitant bit of prognosis, always a risky undertaking.

One point some modern readers may find surprising is the role of government, in general, and the military, in particular, in sponsoring both science and invention. It's not a new phenomenon. It's not necessarily a bad thing. When a government is faced with a crisis, it often looks to science for a solution.

This is a relatively short book for such a large subject. My intent is to interest readers, not to replace the huge amount of material that exists about communication and awaits the avid student willing to explore it. No technical background is assumed on the part of the reader. The book should appeal to anyone from high school age and above who likes reading and understanding the world in which we live.

I have tried to compromise between the seemingly endless citations of a scholarly work and providing the reader with sources for further reading. Often I have a single note for a section with a number of citations, rather than clutter the text with footnote symbols. I have also used metric units for all measures as is standard in scientific writing. The first note in most chapters provides citations to general readings on the subject. In many cases there are brief vignettes of the contributions of famous people or those who made important discoveries, but there is only room for the most significant among them. There are a great many citations to the *Encyclopaedia Britannica*. Since this now occurs in so many forms (print, disk, and online) I have shown only an article title and no details, since it will have different locations in each of these media. A fine comprehensive bibliography on the subject is found in Christopher H. Sterling and George Shiers, *History of Telecommunications Technology, An Annotated Bibliography,* Lanham Md., Scarecrow Press, 2000.

As usual, many people other than the author played a role in the making of this book. Special thanks go to Alison Meadow, who did much of the background research as well as proofreading and Sue Easun, my editor at Scarecrow Press, who kept the faith. Others who read and commented on all or part are, in alphabetic order: Saad Bakry, David Kingsland, Alison Meadow, Mary Louise Meadow, Stephen Meadow, Ellen Sleeter, Debby Stewart, Elaine Toms, Debby Wallace, and the anonymous reviewers selected by Scarecrow Press.

Libraries: the University of Toronto Libraries and individuals too numerous to name, the Royal Ontario Museum, the Diaspora Museum in Jerusalem, the Union Pacific Railroad Museum, the museum of Scotts Bluff National Monument of the National Park Service, the Erie Canal Museum, and the Collingwood Museum.

Individuals who gave me invaluable information: Brian Lang, Panasonic Canada Inc., David Waterhouse and Robert Garrison of the University of Toronto, Chuck Schell, CANAC International Inc., Joyce Rogers, Alan Molitz, Paul Milgram, and all the many other companies and museums who supplied illustrations.

Part 1

The Basics of Communication

Communication is so basic to our existence that its fundamentals may seem either obvious or very pedantic, of interest only to academics. Taking a close look at something we've been doing all our lives can be demanding, like trying to think how we walk while doing it. So, we have to begin our trip through the history of communication with some fundamentals and definitions that, I hope, will make the rest of this narrative more meaningful.

Much of communication is concerned with technology. An alphabet, a system of written symbols, together with a system for combining words into meaningful statements, syntax, are inventions, a form of technology. They do work. They convey messages from one person to another. The printing press is also technology. Its work is recording symbols repetitively so that copies can be transported to readers far from the print shop. A musical instrument is technology, even if it takes an artist to create it and another to play it. The telephone and radio are inventions, machines that do some work. The work is carrying messages from one point to another.

The successive technologies of communication have had profound effects on society. Indeed, for better or worse, we could not have our present society without them. Their history, though, is far more than a recitation of who invented what and when. It is a recitation also of what needs may have inspired an invention and what effect a new means of communication had on people. Today, changes in communication technology are coming at a dazzling pace and we cannot foresee the effects nor even assess them after the fact, before the next big change comes along.

Jean C. Monty, chairman of Bell Canada, said the following about the history of, and recent changes in, communications in a speech to the Chambre de commerce et d'industrie du Québec métropolitain:

> When you think about it, the history of humanity is almost inseparable from that of communications. The ancient Greeks invented signs, the Romans the first newspaper, the Germans the printing press. Between 1457 and 1500 alone, some 20 million works were printed in Europe—a sizable number indeed considering the limited technical means of the period.
>
> · · ·
>
> This revolution is presenting telecommunications companies with major challenges. It's a well known fact that the telephone networks were designed to transmit voice signals and not data. Yet data traffic is growing ten times faster than [voice] telephone communications. The pace is such that, in the year 2000 [this talk was given early in 1999] data will represent 80 percent of the overall traffic on our telephone networks. The growth will force us to be more creative in the solutions we offer our customers and fundamentally change our operating methods.

Starting in part 2, we are going to follow a chronological approach, but not exactly. It's rather a combination of telling the tale roughly in the order in which things happened but occasionally telling about how some particular technology continued to develop, jumping ahead of the story of other, newer ones.

In this introductory part we look at a series of basic questions concerning the nature of communication, the better to understand the context of the new developments of any age.

This is a fascinating story, one that affects every human being.

1

Background[1]

Introduction

Looking back into history prior to the middle of the twentieth century, the three technological developments that brought the greatest social change to our society were the alphabet, the movable-type printing press, and the telephone. There have of course been others: wind and steam-powered ships, railroads, electricity generation and distribution, automobiles, steamships, radio, airplanes, and television. In the middle of the twentieth century came computers, communications satellites, and fiber optic cables. All these can be seen in some sense as communications media.

There is no single measure of the impact of such technologies on society but we might consider how quickly a major societal change followed the invention. Steam locomotives, automobiles, and airplanes each took about half a century to go from invention to a dominant position in travel, which is a form of transmission, but they are expensive machines and their operation required an extensive infrastructure. Television, being a sort of combination of radio and cinema, was not so different from what preceded it, but it brought live visual information and entertainment into the home. Its net effect is still not clear. Gutenberg's printing press and Bell's telephone had effects that were as immediate as anything could be in their times. Each was clearly the right invention for its age.

An indication of the pace of change today is that the telephone took forty years to gain its first ten million customers. Facsimile (fax) required half that time. Personal computers did it in five years and electronic mail but one year.[2]

We are not yet ready to assess the total impact of computers on our lives. Today, after half a century of these machines, we have a convergence of computers and communications called the Internet. A point of terminology: computers talk to each other by using the same basic communication means that telephones do. These include electric wire, coaxial cable, glass fiber, and microwave radio. Computers send different signals than telephones, but over the same kind of facilities. They require the device known as a *modem* to be able to send and receive messages over telephone lines. They are not, themselves, the means of transmission; they are a means of getting the message into the transmission channel.

Many feel the computer is the successor to the printing press and the telephone as a technology ushering in a major round of social changes. *Is it that great a change?* Probably. *Will its effects be major?* They already are. *Beneficial?* Only time will tell, but the potential is there and rate of increase of usage continues to astonish everyone.

A few years ago, an AT&T television advertisement claimed they were the company that invented telecommunication. This would be true if you accept that the telephone was really the beginning of telecommunication because, as we shall see, AT&T is a corporate descendant of Alexander Graham Bell, inventor of the telephone. But if we take the literal meaning that telecommunication means distant communication, we have to go much farther back into our murky past, before written history. Distant communication is at least as old as civilization. It may well have made civilization possible or it may have been the inevitable result of early civilization.

Businesses, schools, hospitals, and government agencies now routinely operate in several countries and service clients all over the world. So do individuals. We can talk by telephone from almost any metropolitan area to any phone in the world. We can use automatic bank teller machines almost anywhere in the world to withdraw money in the local currency from our home bank accounts. We can, if so driven, follow the stock market anywhere the telephone works. We can determine our geographic position anywhere in the world. To do any of these requires effective communication: fast, cheap, and reliable. Today's society cannot operate without the media that make this possible.

The Basic Questions

We will bring out the fundamentals by asking and answering five questions:
1. What is communication?
2. What is information?
3. What is a medium?
4. How is information represented?
5. Was McLuhan right: is the medium the message?

What Is Communication?

It may be surprising that communication, an activity so basic to our existence, has many different definitions and meanings to different people. The simplest way to think of it is that *communication is the transfer of information* (hence the second question). To transfer information means that the information must be carried in some manner from the originator to the destination. Figure 1.1 shows the basic concept, originally presented by Claude Shannon.[3] The means of carriage or conveyance is called a *channel*. An example is a telephone line. Another is the post office letter carrier. To go from the brain of an originator of a message to the brain of a recipient, the message goes through several transformations. A telephone message must be made into sound, by the voice, then into electric current by the telephone instrument in order to be carried by the wire. At the receiving end the electric current is rendered back into sound. The mechanisms that do these transformations are called, generically, *transmitters* and *receivers*. Transmission can span time as well as space.

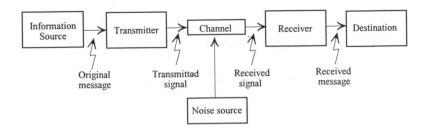

Figure 1.1. *The basic elements of a communication system.* An information source might be a person, the transmitter a telephone instrument, the channel a series of wires and cables interconnecting telephones, the receiver part of a telephone at the destination's end, and the destination another person. Any communication system is subject to some noise being added. Unfortunately, as a result, the received signal or message might not be the same as the original message or signal sent.

For there to be communication, a message must be transmitted *and* it must be received. It's not enough to talk into a telephone or put a letter in a mailbox. The recipient must receive the message. Oh, if the companies that over-use voice mail only understood that!

Figure 1.2 shows a variety of the symbols used in print communication. What kind of effect do they create? Each brings to mind an organization, a belief system, or a means of creating other meaningful symbols. We usually think in terms of a change in the memory or knowledge of the recipient based on the assumed

meaning of the message. Ah, *meaning*! That is what is truly at the
heart of the matter. So, we have information, communication, and
meaning. Let's struggle with these one at a time. Readers wishing
to explore the subject will find other usages, but careful reading
usually shows that though different terminology may be used to
describe communication, information, data, and meaning, most
writers eventually draw the same distinctions among whatever
terms they use.

Figure 1.2. *Symbols used in print communication*. Any symbol can
convey information so long as it is recognized and understood to have the
same meaning by both sender and receiver. Here are some very common
ones. Depending on the culture of the reader, some of these may not have
much meaning.

Today, we hear much about the media, often equating the term
to news or entertainment distribution companies, but there are
other meanings to be explored under question three. We are also
often hearing the media talk about *content*. This is nothing new,
but the term has become an "in" expression. Content is what the
media deliver. It's the words, songs, pictures, ideas. It definitely *is*

different from the media that transmit it but can be much affected by the medium.

What Is Information?

One of its many definitions is that information is the *effect created* in a *recipient* by a *set of symbols*. If you read enough on this subject, you will find that almost every author has at least a slight difference in how these terms are defined.[4] With that understood, let us proceed with our own definitions. *Data is* (or *are* if you're a purist) *a string or set of symbols*. We do not have to understand data. They may be just a series of numbers, perhaps measurements of something that we do not know about or words in an unfamiliar language. *Information is data that has been received and assimilated* into the recipient's store of knowledge. There has been some cognitive processing or understanding of the data. Yes, this means that what is information to you may be merely data to me because each of us may mentally process the data differently, including rejecting it entirely, depending on what was in our respective minds when we received the message.

Information does not have to be true. It just has to have some meaning, some relationship to what we already know. If I tell you a falsehood and you recognize it as such, you have the information that I lied.

Knowledge is the cumulation of information in our minds. It is not just a file of received messages, but the integration of a lifetime's received messages with each other. Just as an individual message may be information but not true, knowledge is not true or false for everyone. Just think a bit about religion, politics, evolution, or the wisdom of the latest baseball player trade.

We hear much about information overload, the curse of the information age. Is it possible to have too much knowledge or understanding? If you believe that ignorance is bliss, your answer could be yes. But today few intelligent people subscribe to that philosophy. Most of us recognize the value of information, yet at times feel overwhelmed by the amount that appears to be out there. This is partly a definitional problem. If information is that which we have received and understood, then there cannot be an overload because overload implies that there is more than we can receive and understand. The solution to this conundrum is in the definition. There is no question about the huge volume of *data* that seems to be everywhere. When we think we have information overload, what we have is an overload of *potential* information, of as yet unappraised or unassimilated data. However much of that load of message traffic we do manage to assimilate becomes information and adds to our knowledge. The part we can't assimilate makes our heads spin.

These are not just idle academic musings. Communication is a very pragmatic activity. We are not usually interested in merely transferring data from one location or time to another. We want to transfer information. We want to be sure that what we send can and will be understood by its recipients.

Meaning

To be understood, to constitute information, a message has to have meaning. Here is where we really run into some semantic problems. "The Giants massacred the Indians" would be horrific to a person who knows English but has no knowledge of American sports, of the way sports teams are named, or that *massacre* may mean only that one team scored many more points than the other.

The message "KO last traded at 81-1/8" is comprehensible at one level—we all know what the individual words and numerals mean, except for the nonverbal symbol KO. But until that symbol is recognized as representing the Coca Cola Company and we know the context is that of the stock market, there is some uncertainty as to what the message means. Uncertainty is the key. One way to measure information is to measure the amount of uncertainty it dispels. If you do not know what KO is, the message dispels no uncertainty. Therefore, the amount of information in a message depends on who receives it and what that person already knows.

The meaning of *meaning* is a deeply philosophical question. Practical communications engineers know that if recipients cannot understand the messages they receive, there is no point sending them. That is a useful thought in all aspects of human communication. How many of us have said about another person, "It's no use talking to X. He doesn't understand?" Or, as my father used to say, "Talk to him, talk to the wall." Morse code did not become a useful device until enough people knew how to send and receive it. Internationally standardized road traffic signs are useful because "everyone" is expected to understand them. Not every driver understands the German warning, *Glatt eis gefahr*, but almost any driver understands a picture of a car skidding on an icy road (figure 1.3). Many professions and trades develop a highly stylized jargon that is incomprehensible to outsiders. Perhaps to some extent this is done to enhance the status of the in-group, but it also tends to make communication among that in-group easier and less ambiguous and meanings clearer. Intentional ambiguity can be introduced in order to confuse a message recipient—something that may sound like it belongs in a modern spy novel but is a very old practice.

GLATT EIS GEFAHR

Figure 1.3. *A graphic road safety sign.* In spite of the possibly unfamiliar language of the words, the message is readily understandable to anyone who drives a car wherever it rains or snows, by means of the pictorial symbol.

Context Dependency

Misunderstandings in communication are all too common and they happen easily. A common cause is that the sender and recipient of a message have different contexts in mind.

What does the word *field* mean? A number of things. It could refer to an open area of ground used for agricultural purposes, an area of military combat, an area used for certain athletic contests, a realm of specialized work or knowledge, or a space influenced by a magnetic force. What does "He's in the field" mean? It means nothing if the recipient does not know which of these contexts is intended.

The statement "I like that one" requires the recipient of this message to know to what "that" referred. *That* is an anaphora, a word that refers to and takes its meaning from some previous word. Without knowing what the referenced word was, we cannot interpret the sentence.

Thus, the context in which a word is used often determines its meaning. When more than one meaning of a word is possible and the choice depends on the context, we call that *context dependency*. A 1998 legal proceeding involving President Clinton brought us such expressions as, "depends on what the meaning of the word *is* is."[5] Out of context, it sounds silly. In context, it seemed to emphasize the difference between "is now" and "has been over time." We shall see in later chapters examples of using context dependency to simplify communication.

These have been examples from *natural language*, a language that we "naturally" speak, that we learn from our parents and others around us. In such languages, especially English, it sometimes seems that everything is context dependent. But, in mathematics and computer programming, which make use of *artificial languages*, symbols may be *context independent*, or have meanings free or almost free of any dependency on context. We shall see communications systems that are heavily context dependent and others that are not.

Communication of Meaning

Whenever we undertake to communicate we are trying to inform, entertain, warn, convince, or arouse someone to action. In other words, telecommunication is purposive. We don't do it for its own sake, we do it to accomplish something, which might only be to make ourselves feel heard or understood.

Any of these goals requires that the originator of a message takes some pains to assure that its recipients understand the content sufficiently well to accomplish the objective and that the means of transmission are appropriate to the message. St. Paul wrote

> And even things without life giving sound, whether pipe or harp, except they give a distinction in the sounds, how shall it be known what is piped or harped?
>
> For if the trumpet give an uncertain sound, who shall prepare himself to the battle?
>
> So likewise ye, except ye utter by the tongue words easy to be understood, how shall it be known what is spoken? for ye shall speak into the air.
>
> There are, it may be, so many kinds of voices in the world, and none of them is without signification.
>
> Therefore if I know not the meaning of the voice, I shall be unto him that speaketh a barbarian, and he that speaketh shall be a barbarian unto me.[6]

Paul did not write in this Elizabethan English of the seventeenth century. These translated words, to the modern reader, may fail his own criterion of being "easy to be understood."

What was meant earlier by "sufficiently well" and "appropriate?" Suppose you want to send this message: "Meet me at 10 o'clock tomorrow." It would be sufficiently understood only by someone who knows where you intend to meet and whether you mean morning or evening. If you are close friends and have an established pattern of a place and time to meet, this is perfectly clear. If this is someone you never met before and you have had no previous communication about places, it is almost meaningless.

"Appropriate" may refer to speed, accuracy, nature of the coding or representation of information, or degree of recipient involvement. In this case, it would not be appropriate to send the message by conventional mail because it usually takes more than a day to get to its destination. Not only would no place have been mentioned, but the meaning of "tomorrow" would be lost.

To transmit racetrack results or the price of shares on the stock market, you would need a fast transmission system, accurate in reporting the name or number of the winning horse or symbol and price of the stock you are describing. You would not have to show a picture of a horse or stock certificate, reproduce the sounds of the race track, or hold the destination's attention for more than a few seconds.

Television conveys a strong feeling for news events, especially disasters such as war or storms. Here, we do need pictures. In these cases neither print nor radio are nearly as effective as graphic images, and motion pictures or stage drama are too far removed in time from the immediate reality to be as effective in conveying the feel of a dramatic current event.

In presenting fiction to an audience, an author has a choice of media: storytelling, printed books, live theater, radio, song, television, or cinema. Each of these media is different. Among other differences, each calls for a different degree of audience involvement or participation. A book without illustrations requires the reader to imagine the scenes and characters. Radio requires a great deal of imagination on the part of the listener, but it can provide realistic sound effects and the voice tones of the actors. Early television asked much of the viewer. Cinema and modern, sharp-image, color television invite the reader to sit back, relax, and let the presentation do all the work. Question 3 explores more about media.

Claude Shannon and Warren Weaver, in their landmark book[7] on information theory saw the problems of communication as occurring at three levels:

1. The *technical level*, concerned with the mechanical aspects of transmitting or conveying symbols from one point to another.

2. The *semantic level*, concerned with meaning. Does the recipient understand what the sender wanted or meant to say? How can the probability of understanding be kept at a high level?

3. The *effectiveness level*, concerned with the effect of the symbols on the recipient. Did the person do what you wanted done? If you shouted, "Help!" did you get help? The *Titanic*, in effect, shouted "Help!" but got very little of it.

What Is a Medium?

We have five senses that bring us information. We have the corresponding ability to send out signals to trigger the five sensory receptors of others. The senses are the physical means by which humans communicate. We can use gestures or write symbols that others can see, use speech or other sounds they can hear, touch them, offer them perfumed gifts or tasty foods. We spend a great deal of our time transferring and receiving information. Without communication human life is almost unthinkable. The loss of even one of our senses creates difficulties, but we have a remarkable ability to compensate by extra development of the others. The best known example of this was Helen Keller who learned to use the sense of touch to overcome her lack of ability to see or hear.[8] This suggests that much communication is redundant, that information can be sent in more than one way. Animals, too, communicate. Wolves, chimpanzees, and bees, for example, communicate among themselves as a necessary part of their lives and it has been suggested that communication occurs even between molecules.[9]

Our abilities to see, hear, and smell can transcend distance to various degrees. We can see over enormous distances to distant galaxies many light-years away. We can hear only relatively local noises, limited to a range of a few hundred meters for the human voice to a few kilometers for thunder. Smell requires that the minute particles that carry odor come into physical contact with us, even though they may have traveled some distance to do so, typically limited to hundreds of meters. Touch requires, well, touching, being in direct contact. The distance information can travel is determined by the nature of the medium that carries it. Light travels as electromagnetic waves through empty space; sound waves require a substantive medium, such as air, to carry them. Both kinds of waves dissipate or attenuate to some extent as they travel; radio waves can penetrate solid walls before losing all their power. Light is easily stopped by a solid medium. Sound falls between these two in its ability to penetrate barriers. We humans can directly sense light but not radio waves. We require them to be converted first into sound or pictures.

Kinds of Media

The words *medium* and *media* are frequently used in communications. Unfortunately, they have multiple meanings. One is that a medium is the physical means of conveying a message from origin to destination—the electric wire of a telephone system, the air in which sound waves travel.

A variation on this is medium as in *medium of the press*. Here, we are not literally talking about the machine that puts ink on paper. We are talking about an organization or enterprise, or many organizations that, collectively, constitute the newspaper industry. Television as a medium can have either meaning. We can think of it as the bundle of technology that carries pictures and sound to us from afar, or as the production and distribution industry. The Pony Express was a medium in this organizational sense, as is a school or publishing company.

A medium can be the physical material upon which a message is recorded, not entirely different from the first definition, but traditionally applied to recording, not transmission of messages. An artist can work in the medium of clay or of oil paint. This book is recorded on the medium of paper, but *paper* can also refer to the institution that produces a newspaper or to the complete, printed journal they produce.

Finally, a less common meaning is a person who carries messages. One such is a spiritual medium, conveying messages between the earthly and spiritual worlds. But the medium can be anyone who serves as a go-between.

The Media of Communication

Before writing, there was no way that a person in one place at one time could communicate with another person at a different time and far-distant place. This was the great advantage writing gave us. Prior to writing, if a message was to be saved it had to be saved in a fallible human memory. The great significance of writing was that for the first time it permitted communication *over time*. Something could be written or drawn today and read tomorrow. This was not possible with the voice, drum, or smoke signals. The earliest written history, that is texts written *as* history, dated from about 2500 BCE.[10]

Early writing was closely related to drawing, painting, and carving. Modern multimedia seem to be returning us to this state of affairs. Music and dance (a form of gesture) are also very early media. There is no well-established date when the first human recorded the first graphic image. Figure 1.4 shows a drawing made by a modern archaeologist of figures found in a cave in La Marche, France, dating from about 20-40,000 years ago. The main figure is a pregnant mare with darts or spears thrust into her side, believed to represent a ritual killing. The smaller, geometric symbols seem to be tallies of days, probably to mark the time when the ritual should be carried out.[11] There is no indication that this form of communication directly became the basis for later writing. The respective cultures that developed these forms were nowhere near each other.

We know a fair amount about how writing started because the media upon which early written symbols were recorded was often clay, which hardened into something long-lasting like stone. Writing is generally believed to have started in the remarkable land of Sumer, in what is now Iraq, as early as 8000-3000 BCE.[12] Sumerians also developed the wheel, the arch, crop irrigation, and written history and law.[13]

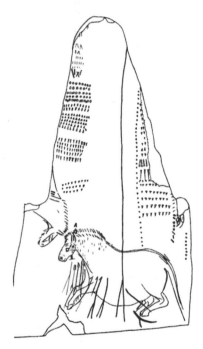

Figure 1.4. *An early cave drawing.* This is a sketch made from a piece of bone believed to date from 20-40,000 years ago. The various geometric figures—circles and triangles—probably indicate astronomical phenomena such as phases of the moon. The darts striking the pregnant mare are probably indicative of sacrifice rituals. Drawing courtesy of Alexander Marshack.

Early writing was largely commercial and enabled trade among distant peoples. Commerce spread culture, including the knowledge of how to write and, of course, read. The spread of culture was not always a pleasant, refined activity like going to a foreign movie. It was often quite brutal, as in the overrunning of Rome by Central European hordes or of the Americas by Europeans. But it was communication nevertheless.

Between the development of writing and the development of electrical transmission in the mid-nineteenth century, communication mostly advanced through improvements in recording media and transportation. When Egyptians learned to use papyrus, a plant growing in the Nile River, to make a medium for writing, they had a thinner, lighter medium than clay or animal hides. People could not shout a message across the Mediterranean Sea or the Alps. They had to write down the message and have someone or something carry it to its recipient. The soldier who brought the news of the battle of Marathon to Athens in 490 BCE is still celebrated.

The Pony Express was a mail service. The railroads carried not only passengers and freight but mail as well and they were early users of the electric telegraph. Air mail (figure 1.5) provided much of the economic incentive for the development of commercial aviation.

In 1844 Samuel F. B. Morse demonstrated his electrical telegraph, and the field of communications has been changing, almost wildly, ever since. What Morse did, not for the first time but for the first time in the United States on a practical scale, was to enable the sending of messages over long distances, seemingly instantaneously, without the need for transportation of the written message. This is certainly *telecommunication*, far or distant communication. Today, the electronic media, radio and television, have almost eliminated distance as a major factor in communication.

Figure 1.5. *The beginnings of airmail.* Shown here is an airplane about to be loaded with mail for one of the first U.S. air mail runs in 1918. Airmail speeded delivery and helped develop the aviation industry. Photo courtesy National Archives and Records Service.

Media—Hot and Cold

Marshall McLuhan classified media as hot or cool.[14] A cool medium is one of low definition (like the telephone) calling for the recipient to become involved, to do much of the work of understanding the message by mentally filling in what is not explicitly presented. A hot medium, such as cinema, is one of high definition, requiring less effort by the recipient to complete the images. It does all or most of the work for us. We sit back and become engulfed by it. McLuhan saw radio as hot, television as cool. The black and white television of the 1940s and 1950s and the early color sets were certainly cool, but modern television, especially the newest high definition television, is more cinema-like, presenting lifelike images and requiring little of the viewer. Deciding whether to send a message via a hot or cool medium is a choice an author or originator must make. Which medium is best? There is no fixed answer to that question. Best for what purpose? Best for whom? Much depends on how information is shown.

How Is Information Represented?

A message has content. It is supposed to convey information. How do we represent the information in the message? How do painters get their message across? How do writers? How does a telegraph operator convert a handwritten message into a form that can go over the telegraph wires? We will discuss four aspects of information representation: the symbols chosen to represent an entity or concept, direct and indirect representation, the amount of variation expected in symbols used, and the difference between analog and digital representation of symbols.

Choice of Symbols

Much of what is involved in the choice of symbols to be used for a message has to do with content rather than the means of transmission. The composer of background music for a motion picture will use certain types of music to enhance a sense of foreboding or happiness. The costume director in old-time western movies might dress the hero in a white hat and the "bad guys" in black. But some choices have very much to do with how messages are transmitted. We shall see in later chapters that early ideas for the telegraph involved rather cumbersome ways of representing letters and these would affect the practicability of the system. Morse's dot-dash code was efficient but it required a skilled operator. There is continuing debate in France and the Canadian Province of Québec, for example, about whether voice radio conversations for air traffic-control should be in English or the language of the region in which controllers are located. At issue are both national pride and the probability of being understood.

Communication systems must take into account the means of representation that users will want to use.

Direct and Indirect

A. R. Taylor[15] recognized two types of symbol systems, direct and indirect. In *direct representation* the symbols are directly recognizable, for instance, a picture of a bus on a bus-stop sign, the word "arf" to represent the sound of a dog barking, or a rectangular figure to represent a house. Figure 1.6 shows the development of two letters of our alphabet from their earliest form as pictographs. The Hebrew letter *bet*, ‏ב‎, evolved into the Greek *beta*, β, and then to our B, which was derived from an earlier Phoenician letter and that from an Egyptian rectangular pictograph representing a house. *Bet* or *beth* still means house in Hebrew and the symbol bears some resemblance to a picture of a house. Our M is believed to come originally from a Sumerian symbol for water, similar to an Egyptian one. Neither letter has a semantic meaning to us. They now represent only sounds.

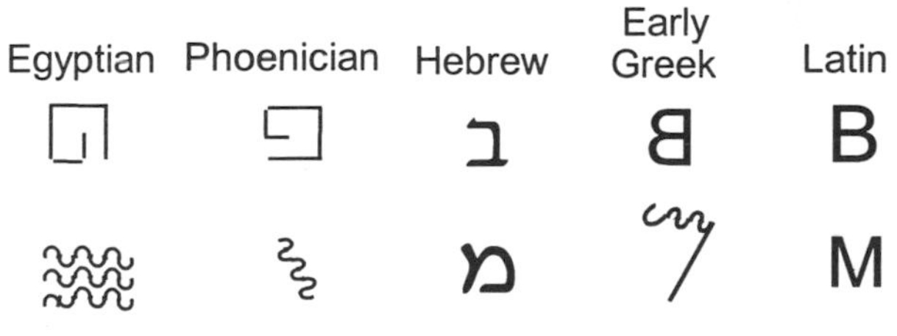

Egyptian	Phoenician	Hebrew	Early Greek	Latin
⌷	⊐	‏ב‎	ᗺ	B
〰	⋛	‏מ‎	ᔓ	M

Figure 1.6. *The evolution of some letters.* Here are an Egyptian hieroglyphic representation of house and water and corresponding letters of the Phoenician, Hebrew, Greek, and Latin alphabets that derived from the early form. Is the Hebrew *mem* (‏מ‎) related to the Phoenician or early Greek symbols? It's not clear.

In *indirect representation* what is represented is some aspect of the language of the users, not the concepts to be communicated. Morse code symbols can represent sounds that in turn represent letters, which in combination represent words. Direct systems make it easier to communicate between people who speak different languages. The universal signal for asking a waiter for the bill is to make the motion of writing on your hand. Hand traffic-control signals used by police are at least close to universal, and we now have international standards for motor vehicle traffic-control signals.

Umberto Eco[16] tells of the medieval belief, coming from *Genesis* 2:19, that "Whatsoever Adam called every living creature, that was the name thereof." Scholars of the time believed that words of the language spoken by Adam were the *right* or *appropriate* words for the objects or beings referred to. Our word *zebra* sounds arbitrary. In English there is nothing about the word to suggest the animal. But by this theory whatever Adam called that animal was the word that *should have* been used, the word would bear some relationship to the animal. If so, that surely would have been direct representation. Our word *rhinoceros* does not appear to be direct, but it comes from two Greek words for nose and horn. (figure 1.7) For a Greek speaker, then, *rhinoceros* is direct or certainly more so than it is for the rest of us.

Figure 1.7. *The name on the nose.* One example of how objects, or creatures, get their names. The rhinoceros is named after the Greek words for nose and horn, hence in a sense is the name it *ought* to have been. Photo © 2001 www. arttoday.com.

Variation

Another important aspect of symbols is the extent to which they are permitted to vary, an important part of which is the extent to which any given symbol can be *expected*. To take an extreme example, if a mason carves the name of a building in a marble lintel over the door, that name is always there. It is hardly news that the name is still there the next day or even the next century. There was no need to plan for name changes when erecting the building. When two teams play a ball game, only one of them can win. So there is always some question about which one it will be. Therefore, in a report of the result it is not enough to say the score was 3-2. The report must also make it clear which team won. If we roll a pair of dice, we know the total of the numbers showing must be in the range 2 to 12, but we don't know that total in advance, so there is always some element of surprise.

In these three examples, the symbols—name of the building, name of a winning team, number showing on the dice—show different degrees of variation. Such instruments as a door buzzer or a ship's whistle can send only one sound. All that can vary is the number of times the sound is made, the duration of the sound, or time between successive sounds. The difference between a telegraph machine, using sound to represent dots and dashes, and a telephone is that the phone can transmit a wide range of tones, the telegraph only one. A transmitting device that can handle more variation can send more information in a given amount of time.

Analog and Digital

By analog representation of a measure, we mean the use of one measure to stand for another. Time measurement provides some well-known examples. A sun dial uses the position of a shadow to represent the time of day or the location of the sun in the heavens relative to earth. At one time candles were used as clocks, the candle being made with alternating layers of color and, since it burned at a uniform rate, the passage of time could be measured by the number of layers that disappeared. Finally, an ordinary clock or watch uses the position and movement of the hands as an analog of the passage of time. Modern digital clocks display numbers. A clock showing 1:34 means that thirty-four minutes have elapsed since it was one o'clock. Here, the numbers are shown directly.

Figure 1.8 shows a modern watch face with a combination of analog and digital representations: the time is analog, the day and date displays are digital.

Some computers, now relatively rare, used the angle at which a shaft was turned, the intensity of electric current, or strength of a magnetic field as an analog of a number to be stored. In their digital counterparts, numbers were represented by symbols such as 1010, the binary number system representation for the decimal number 10. (The binary digits indicate, from left to right, one 8, no 4s, one 2, no 1s. Add them up and we get the decimal number 10.) Today, analog computers survive only in highly specialized applications

Figure 1.8. *A watch face combining analog and digital representations.* The hands are analog devices. The angle of any hand is an analog of the amount of time that has elapsed since it was at an arbitrary starting position, usually pointing straight up. The day of the week and date are shown in digital form. Photo by C. Meadow.

In sound and photographic or video recording, analog representation lasted much longer and is still used, but digital recordings are rapidly gaining favor. What we are finding is that digital representation gives more accuracy, hence a digital musical recording or broadcast can give a truer representation of the original sound than analog. In pictorial recording and transmission, digital representation also prevents loss of accuracy in transmission as well as giving us the opportunity to manipulate images by digital computers. We now have digital television, digital voice transmission by telephone, and digital cameras, as well as clocks and computers.

Was McLuhan Right?—Is the Medium the Message?

In order to choose the appropriate coding and transmission medium for a message, the sender must think about how it will be received. If you are trying to arouse the patriotic fervor of a group of people, say to get them to support their government in a war or their school's football team in a game, is this best done with a printed text or in a live meeting, with many people, a dynamic speaker, a band playing, and uniformed soldiers or players looking very heroic? The actual words conveyed by the different methods may be the same. The impact will be vastly different.

Drama is another example where there is a choice of media. We pointed out how different a drama would be on radio compared with cinema or television. It is also different in live theater. A motion picture or television recording of a live-theater presentation does not make for good movie or TV viewing. The sets rarely look real, something we expect from cinema unless it is fantasy or animation. Recent advances in the technology of film animation enable the artists to make more realistic movements than the jerky stuff of Tom and Jerry cartoons. That jerkiness creates a comic feeling. Smoother motion enables animated films to be more realistic or dramatic. Disney Studios' *The Hunchback of Notre Dame*, *The Lion King*, and *Dinosaurs* are examples. These are not just slapstick comedies. They can express serious emotion because the pictures are more detailed, hence realistic or impressionistic, and the motion is smoother, hence more realistic.

The number of different settings in a stage presentation must be limited and the props must be easily portable, while in the movies there may be many sets, some real and quite immovable. The very way actors speak is different in the two media. The styles of speaking are not interchangeable. The movie actor does not have to project to the back of the theater but he can whisper and still be heard. The book and theater versions of a story demand a high degree of imagination of the reader or watcher, building the mental image of a real-world setting from the written descriptions or the minimal stage sets. The television and cinema settings do it for us. A book can describe a scene as slowly as its author wants. On the other hand, television is known for presenting everything in a series of short, high intensity bursts, that keep our attention but ask little of us.

In his book on baseball broadcasting, Curt Smith[17] says that radio is the perfect medium for broadcasting baseball. Actually, it depends on the listener. If someone knows the game well, the radio broadcast can conjure up all the images needed. The listener's imagination does the rest. For someone who does not know the game, seeing what is happening will surely convey a better understanding. The television view shows less of the field than a person attending the actual game sees. Again, if the viewer does not know the game reasonably well, the broadcast images may be meaningless since the televiewer sees one or just a few players at one time but sees them close-up, in detail. Those of us sitting in the cheap seats see the whole field of action but little detail. None of the people at the live game are offered explanations of unusual occurrences on the field, but the broadcast audience does get this and from an expert observer. Hence, many in the audience will bring along a portable radio and listen to the game while watching it live. That way they both see the action and hear an expert commentary by radio.

In both still photography and motion pictures there is a difference between black and white and color images, but which is better in any situation depends on what the photographer or director is trying to say. Ansel Adams's dramatic photos of Yosemite Park need no color to enhance them. When computer-based colorization of black and white motion pictures became possible it may have enhanced their value to television but it enraged some of the film makers.[18] These are different media and they convey a message differently. Poor colorization detracts from the moods intended by the authors and directors.

McLuhan's most famous observation was that "the medium is the message."[19] We take this view nonliterally. Media certainly affect content. There can be no question that the medium *affects* interpretation of the message. The example was given of choice of media when arousing a football crowd. But, is the medium, literally, the same as the message? Take a very pedestrian example. One message is "Meet at 1 PM." Another is "Meet at 2 PM." If sent by the same medium, are these the same? Hardly. But, if one is sent by postcard and the other by a skywriting airplane, their effects are likely to be very different. Sending a personal message by airplane, for all to see, sends far more than just the content. It also sends the message that the sender really cares, really wants the recipient to get excited at the prospect. Skywriting or conveying a message written on a banner towed by a low-flying airplane has been used as a means of sending a proposal of marriage (figure 1.9). That is impact.

As we proceed now through the history of communication we will see that technology changes often and dramatically but the principles do not. Although we occasionally mention animal or machine-to-machine communication, we are basically concerned with *human* communication, typically from one person to another or to many, and often considerably augmented by technology.

Figure 1.9. *Conveying the message with impact.* When a proposal of marriage is sent this way it has impact. This is a reconstruction of an actual message the author saw delivered in San Diego, California, in 1999.

Figure 1.10. *Communication in the street.* Visible here are wires and cables for telephone, television, and electric power (which enables us to see many forms of communication), electric-light traffic signals, street lights enabling people to see what is happening around them, graphic images used to control traffic, and words used to augment the graphic symbols. Photo by Ben Meadow.

Notes

1. Some recommended general histories of communication are: Crowley and Heyer, *Communication*; Innis, *Bias*; Innis, *Empire*; Oslin, *Story*; Rowland, *Spirit*; Singer et al., *History;* Solymar, *Getting the Message;* "Communication," *Encyclopaedia Briticannica*.
2. Specter, "Your mail."
3. Shannon and Weaver, *Mathematical Theory*, 5.
4 Meadow and Yuan, "Measuring" (A somewhat technical article discussing various definitions of data, information, knowledge, and related terms.)
5. "The President's Grand Jury."
6. 1 Corinthians, 14, 7-11.
7. Shannon and Weaver, *Mathematical Theory*, 114.
8. Herrman, *Helen Keller* and, more generally, Suedfeld, "Isolation" (scientific article on sensory deprivation).
9. No one expects animals to communicate at the same intellectual level as humans, but that they do communicate among themselves has long been observed. One such account is in Charles Darwin, *Expression*. More recently: Busnel, *Acoustic Behaviour;* Horn, "Speech acts" and "Animal Behavior," *Encyclopaedia Briticannica*. Some aspects of dog communication are dealt with in the very readable Budiansky, *The Truth about Dogs*. The view from the cell level is Loewenstein, *Touchstone*.
10. Kramer, *Sumerians*, 34.
11. Marshack, *Roots*, 194
12. Schmandt-Besserat, "Earliest Precursor."
13. Kramer, *Sumerians,* 289-290.
14. McLuhan, *Understanding*, 22-32.
15. Taylor, "Nonverbal."
16. Eco, *Serendipities*, 24.
17. .Smith, *Storytellers*.
18. "Spielberg's Lament."
19. McLuhan, *Understanding Media*, 7-21; Logan, *Alphabet*, 227-247.

Part 2

Telecommunication Before
Steam and Electricity

In this part, we begin at the beginning of human communication. Of course, we don't really know exactly when that happened, especially since there would have been communication of sorts even when our ancestors were not much different from the other animals. Humans learned to communicate, presumably first by gesture and nonverbal sounds, then by words. They learned to draw much earlier than they learned to write, in fact earlier than they learned agriculture and the construction of cities.

But learn to write they did, and they also learned to make paper or substances like paper and then to print by use of a machine. From the time of the invention of printing with movable type to that of the electric telegraph, most advances in distance communication among people were made by improving transportation. In this part we consider all these developments prior to the time when steam and electricity came into practical use. These two developments came fairly close together in the late eighteenth and early nineteenth centuries.

Writing and printing, sound, the use of animals to carry messages or messengers, and the use of nonrecorded visual signals

31

(gestures, flags, etc.) are all quite ancient and are all still in use, in some form. That is one of the lessons we shall learn—that as new communications media come into use they rarely completely displace earlier ones, although they do induce changes. The second major lesson is that fundamentals never change: communication requires not only that a message be sent, but that it be received and understood, that transmission must be reliable, that the proper representation must be chosen, and that appropriate media be used to assure successful transfer of information.

2

Spoken Language and Sound Transmission[1]

Introduction

We do not know exactly how human speech developed. Noam Chomsky believes there is a universal grammar more or less built into all humans so that no matter what particular language we speak we are genetically designed for language.[2] Michael Corballis believes that our spoken language evolved, at least in part, from gestures used by early ancestors of humans.[3] Ezra Zubrow and others believe speech evolved from music.[4] But whatever theories governed the evolution of language, we all use it. If some language transmission or reception facilities are blocked or damaged, such as loss of hearing or sight or speech, we tend to develop alternative forms as Helen Keller did. Figure 2.1 shows another form of adaptation, the use of American Sign Language hand signals for those who cannot hear.

When our ancestors lived in small, nomadic groups, probably all members knew each other and had little need for long distance communication with others living far away. But, as they and language evolved, splinter elements that moved to better hunting or gathering locations might have wanted to maintain some com-

munication with the parent group. There would also have been the
need for cooperation in defending against marauding strangers and
in hunting or agriculture. This early intergroup communication
would have been by speech or gesture. Sending a message to a
distant group would have required that someone go to them and tell
or pantomime the message, or that sound be used to carry the
message over distance.

Can't, Cannot, Impossible   Never

Won't, Will Not, Refuse   Have?, Did? Finish, Complete

Figure 2.1. *Symbols from American Sign Language.* ASL is an alternative
to oral language for those who cannot hear. Reproduced courtesy of
Bantam Dell Publishing Group.

Travel by foot, on the backs of animals, or by boat would have
been the only available modes of communication over long distances. As people became ever more sophisticated, they acquired
a need to communicate across time as well as distance. That meant
creating a message at one time that was to be received at a later

time. This was done by writing, drawing, or carving, which were essentially the same thing. With the understanding that we cannot know the complete history of prewritten communication, let us start with what we can surmise of language development, and then, in the next chapter, go on to writing and other forms of recording.

Spoken Language

There are those who believe that animals can communicate by making and interpreting gestures and sounds, and there is increasing scientific evidence of how this is done.[5] Some people do not like to think that animals can do any sort of symbolic communication. Generally, if you have ever owned a pet, you're a believer. Granted, no dog or cat of mine ever discussed the merits of the neighborhood fauna or the quality of the cuisine, but they could indicate hunger, a disdain for the food offered, a desire to go out, a threat to attack, or a reaction to a threat. Since our brains and vocal apparatus are superior to those of lower animals, even to other primates, our ancestors could make and understand far more sounds than could the other animals, presumably language started with grunts, perhaps coordinated with gestures. Those who believe in music as the precursor of speech believe that song, or rhythmic vocal sounds, led to the evolution of the more complex vocal systems in our bodies that are needed for speech. Such theories cannot be proven or disproven, but at least to the lay reader, they make sense.

Spoken language is ethereal. It can traverse possibly as much as one or two hundred meters of space but leaves no residue. The unamplified human voice, let us say of a football quarterback or an officer in charge of a military parade, can carry that distance if unimpeded by trees or hills. The intended recipient gets the message as it is spoken or not all. To *tele*communicate by spoken language, much beyond that hundred or so meters, requires travel by the speaker, some mechanical enhancement of the voice, or a series of relays.

Some of the words we speak are *direct*, they sound like what they represent, as *plop*, *purr*, or *tintinabulation* or they look like what they represent, as noted previously with A, β, B, and *house*. To some extent even individual sounds are context dependent. English speaking people hear a duck say *quack*. French speakers hear the same duck say *coin* (figure 2.2). But most of our words bear no relation to the thing or concept they represent. They are a code and those who wish to communicate must learn the code in order to exchange information.

Figure 2.2 *Duck talk.* What the ducks are "saying" is contextually dependent on what language the human who hears them was taught to hear and speak. The duck on the left "speaks" English, the one on the right, French. Photo by C. Meadow.

Modern spoken languages are marvelously complex instruments for communicating. Different languages or language groups have different grammatical structures and, of course, vocabularies. Yet, in all cultures children begin to use language around the age of one and are typically quite adept by age three.[6] What's more, in spite of the many rules of grammar, we can comprehend statements even when they violate the rules. Modern linguists generally agree that there are no complete sets of rules for natural languages. Native speakers of English can easily understand such nongrammatical utterances as, "There ain't no more left." Recently, it has even become acceptable to brazenly split an infinitive,[7] a faux pas a generation ago in spite of its frequency of occurrence. We are also not supposed to end a sentence with a preposition, but Winston Churchill is credited with saying about that usage, "This is something up with which I will not put."

The strengths of spoken language as a communication medium are its ease of learning, in spite of its complexity (because humans are essentially wired to do so?); its flexibility that allows us to express a thought in many different ways, an essential for poetry, a good novel, or an explanation of an original idea; and its lack of a need for any mechanical tool except what is built into our own bodies. The major disadvantages are that distance is limited and flexibility can lead to a great deal of misunderstanding as well as beauty of expression. Spoken language is very much context dependent. Understanding an utterance may require a common cultural background of speaker and hearer and may be influenced by facial expressions, bodily movements by the speaker, or tone of voice. Hence, in addition to learning the basics of the language, we have to learn how to express ourselves in different environments and through different media. Writing a text, delivering it as a speech before a live audience, and speaking on the radio or on television all may convey different meanings of the same words. Was this the basis of McLuhan's medium is the message aphorism?

Sound-Based Technology

Speaking is not the only way humans can make sounds. We can also employ tools such as various kinds of drums, horns, or whistles. For example, the limited range of the human voice was too limiting for the native people of the North American plains, who were widely separated, or those living in dense forests or rugged hill country, where natural obstacles inhibited sound. The major pre-electric sound-based communication devices were drums or horns of one form or another. These have been found all over the world and go far back into antiquity.

Flint Instruments [8]

Some flint objects have been found that date from around 40,000 years ago and appear to have been musical instruments. They do not make a loud noise so probably were not used for telecommunication of verbal messages. But tunes or recognizable sequences of notes can be messages. Not surprisingly, these instruments did not evolve into major communications media. But, as noted earlier, interest in music may have been what led to interest in speech.

Drums

Francis Moore, reporting in 1738 on a journey to West Africa, wrote that "[the villagers] have a large thing like a Drum, called a *Tantong*, which they beat only on the approach of an Enemy, or some very extraordinary Occasion, to call the neighbouring towns to their assistance."[9] Generally, low frequency sounds will travel better through this kind of terrain. A typical drum emits low frequency sound which can carry much further than the voice, generally in the range of five or six miles, occasionally up to twenty miles.[10]

H. M. Stanley, of "Dr. Livingston, I presume" fame, wrote in 1878 of the baEna people of what is now the Democratic Republic of the Congo (formerly Zaire) that "the islanders have not yet adopted electric signals but possess, however, a system of communication quite as effective. Their huge drums by being struck in different parts convey language as clear to the initiated as voiced speech."[11] The electric signals referred to would have been the telegraph, the only electrical signaling system then available.

Carrington tells of an incident in the old Belgian Congo in 1877.

An officer commissioned with the task of quelling a revolt among the natives . . . had some knowledge of the telegraphic language used by the indigenous peoples. One evening he took, by surprise, a large talking drum and beat out again and again, *luiza quo, luiza quo* . . . come here, come here. The natives, believing the white man to be some distance away, presented themselves full of confidence and were roped in one by one as they came. Thus it was that all hostilities were terminated peacefully.[12]

It is doubtful that the captured people thought of this incident as ending all that peacefully. We see here not only a use of drums to communicate verbal messages but the use of subterfuge in communication, which we shall meet yet again.

In sound-based communication the ability to transmit tonal variation rather than simple sound or no-sound signals is of great importance. The report of a gun is an example of a nonvarying sound. It is often used for signaling, as in a race to tell the runners when to start and also the timer when to start the clock. It does not matter exactly how the gun sounds, almost any sort of *bang* suffices.

Some percussive musical instruments provide only a single tone: drums (not all), the triangle, cymbals. Tonal variation requires a different instrument. This may be done by having a number of different uni-tonal pieces, as in a xylophone or glockenspiel.

Some drums permit limited tonal variation. A drumhead struck near its center gives a lower frequency tone than if struck near the rim. Today, steel drums provide a much wider tonal range than might have been expected, compared to a conventional drum. Steel drums are a form of tuned percussion instruments.[13] Their construction is illustrated in figure 2.3. We shall meet this distinction again in the history of both telephone and radio—the difference between sending a single tone and a full range of tones.

It is worth considering how the drum communication that Stanley heard was done. The systems used for encoding information on these instruments were quite in keeping with much more modern communications systems.

Figure 2.3. *The head of a steel drum.* The drum is typically made from a 50-gallon steel barrel. Shaping selected portions of the head differently, by hammering, creates an instrument capable of rendering a wide variety of tones when struck. Reproduced courtesy of Tom Dunne.

Some Central African spoken languages are tonal. The meaning of a word or symbol or set of syllables will vary depending on how various parts of the word are stressed. We have this in English to a limited extent: **con**·tent has a different meaning than con·**tent** and **add**·ress is different from add·**ress**. Although each of these pairs of words shares a common Latin root, as used today the connections between them are tenuous. But in some African and Asian languages tonal differences are far more common and important. Thus, it is necessary to express the tonal pattern in conversation, as well as to choose the appropriate words. In some languages there are two tones, high and low, and many drums are capable of producing two tones. In "drum language" only the tonal variation is capable of being expressed, not the words. The hearer must deduce from these tonal patterns which words were intended. Some drums could make but two tones. Sometimes two uni-tonal drums were used. In any case, drumbeats represent syllables—not letters or complete words or the concepts the words describe. They use indirect representation. Here are some examples from Carrington:[14]

li·**a**·la, stressed ↓ ↑ ↓, means *fiancée.*
li·a·la, stressed ↓ ↓ ↓, means *rubbish-pit.*

a·**ti** means *has*
la means *no*
nyan·**go** means *mother* (*nyan* is a single syllable)
san·go means *father*
wa·na means *child*

wana ati la sango la nyango means *child has no father no
   mother.*
This is drummed as: **wa**·na a·**ti** la **san·go** la nyan·**go** or
↑↓↓↑↓↑↑↑↓↓↑.

How does this kind of coding differ from something like Morse
code that was used in telegraphy? The drum codes are not unique.
They are highly context dependent. The recipient must understand
the probable context of the message. A Morse operator receiving
a message does not bear this responsibility. The Morse symbol • —
stands for the letter A regardless of context. Actually, in all forms
of communication when the parties can rely upon each other to
understand context, the physical transmission can be easier. It must
have been far easier to train a person to be a drummer than a
telegraph operator because the drum tones were the same as the
spoken tones; there were no new codes to learn. Both Carrington
and Baldwin tell us that most African drum messages consisted of
stock words used almost daily by drummers.[15]

Wind Instruments[16]

Horns, which make a sound when air is blown through or across a
tube, were also common early instruments. In the Alps, a very long
horn called the *alpenhorn* (figure 2.4), about the height of a tall
man, was used to communicate across the deep valleys of this
mountainous region. A similar horn has been used in Tibet. A form
of flute has been found in European caves, dating from 30,000
years ago and one found in China has been dated from 11,000
years ago. Elsewhere bagpipes serve the communications role.

Figure 2.4. *An alpenhorn*. This huge horn generates low frequency sounds that carry a long distance through rugged alpine terrain. Photo © 2001 www.arttoday.com.

Bagpipes originated in southern Europe but today tend to be identified primarily with Scotland, England, Ireland, and places where people from these countries emigrated. They are used as musical instruments and also as a means of communication. In fact, music in general is a means of communication. British armies as late as World War II were sometimes led into battle by pipers.

Figure 2.5 shows a World War II piper. A whole generation of Americans grew up watching movies depicting the U.S. Cavalry charging at the foe upon a signal from a bugler. The signal was not just any sound, it was a particular tune. We have similar examples from many cultures. The point is that these instruments made a sound that carried well, that could be heard and understood above the noise of battle, and that had to present a pattern (tune) meaningful to the recipients.

The well-known song *Danny Boy* says, in part

> Oh! Danny Boy, the pipes, the pipes are calling
> From glen to glen, and down the mountain side.
> The summer's gone and all the roses falling,
> It's you, it's you must go and I must bide.[17]

This surely tells of a call to arms in Ireland. It was written in 1911 by Englishman Fred Weatherly, then revised to be set to the Irish tune *Londonderry Air*. Although Weatherly's memoirs do not tell exactly what he intended, these were tumultuous times in Ireland. It would have made sense to have used the pipes, a local instrument with a high pitch, to send far-flung signals to the youth of the area, calling them to battle. It did not require words. The music would have been enough.

Figure 2.5. *Bagpipes on the battlefield.* British forces as late as World War II were sometimes led into battle by a piper who created an audible signal and also served as a symbol of a British tradition. While these troops, during the 1942 battle of El Alamein, were not under fire when the picture was taken, the pipers did lead the way in combat. Photo courtesy Imperial War Museum (E.21591).

A similar use of music to communicate comes from Scottish legend. In 1547 there was a conflict between the Campbells and the MacDonalds. The Campbells managed to capture a MacDonald castle whose military commander, Coll Ciotach, was away seeking reinforcements. At one point the MacDonalds' piper, now a captive, saw Coll's boat returning and wanted to warn him that the castle was in enemy hands. By virtue of his profession, the piper had been treated with respect by his captors. He asked for and was granted permission to play a piece, standing on a castle wall, but what he played was not part of his regular repertoire. Coll, hearing this in his boat, assumed the playing of a new tune to have some special significance, guessed at what it was, and turned away, saving himself. Here was another instance of use of deception in communication. The Campbell captors, when they realized what had happened, cut off the piper's fingers. Rather gory, but the story brings out an important communication principle. What Coll heard was *unexpected*. What is unexpected carries *significance.* He had only to guess what was the most significant message his piper might have wanted to send. So, the *absence* of a conventional message was the message.[18]

We have seen that important messages can be sent by means of the primitive transmission technology of ancient peoples. The earliest sound transmitters were probably untuned drums, followed soon by tuned drums and horns of various kinds, if we may take the liberty of calling a bagpipe a horn.

Another very important concept brought out by drums and pipes is context dependency which requires that the recipient understand the context of a message in order to understand the symbols used. Tuned instruments demand more of their users but enable more complex messages to be sent over simple apparatus.

The systems we have reviewed were about as good as any available up to the harnessing of electricity in the late nineteenth century.

Notes

1. Hauser, *Evolution;* Pinker, *Language Instinct;* Sommerfelt, *"Speech and Language."*
2. Chomsky, Language, 35-54.
3. Corballis, "Gestural Origins."
4. Cross "Is Music the Most Important;" Vaneechoutte & Skoyles, "The Memetic Origin"; Zubrow, "Archeologist Investigates."
5. *"Animal Communication." Encyclopaedia Britannica.*
6. Pinker, *Language Instinct*, 262-296.
7. *American Heritage,* 1740.
8. Zubrow, "Archeologist Investigates."
9. Moore, *Travels*, 109-110.
10. Carrington, *Talking Drums*, 29, 30.
11. Quoted in Carrington, *op. cit.*, 8.
12. Carrington, *Talking Drums*, 9.
13. Murr and Tello, "Connecting Materials."
14. Carrington, *Talking Drums*, 32.
15. Baldwin, *Talking Drums,* 38-42.
16. Zubrow, "Archeologist Investigates."
17. Weatherly, *Piano*, 279.
18. Mackay, *Collection*, 11-12.

3

Writing and Printing[1]

Introduction

The term *writing* as it is going to be used here means recording information in some symbolic form, to be perceived visually. This includes handwriting, typing, printing, and word processing. It also includes drawing and painting. The symbols used in any of these media may be letters, numerals, or punctuation marks, but also images of trees, ocean waves, people, or musical notes to be played by a performer. They all require the translation of thoughts or perceived images into a recordable form—letters and words, representational painting, ideographs or pictographs, musical notation, even nonrepresentational art. Nonrepresentational art, as well as much poetry and fiction, does not have a common meaning to all viewers or readers, hence many cannot decide what the originator intended to "say." A viewer is required to interpret an artist's work, usually without any sort of guide or code book. In Marcel Duchamp's famous painting, *Nude Descending Staircase* (figure 3.1), the staircase is readily apparent, but not all would recognize the figure as such. As well, the symbolism in James Joyce's novel *Ulysses* has left many a reader baffled as to his meaning.

Figure 3.1. *Nude Descending a Staircase.* This famous painting in the cubist style, is not obviously seen by everyone as a picture of a nude person. It is a nonstandard symbol. Context is required for understanding. Painting by Marcel Duchamp, *Nude Descending a Staircase, No. 2*, Philadelphia Museum of Art, Louise and Walter Arensberg Collection.

Should all graphic art be considered a form of writing? Artists and critics disagree on this point, but a sizable faction feel that art should be a thing unto itself, not necessarily representing anything. Certainly a picture or musical work does not have to correspond to a text or story to gain general acceptance. But writing was invented as drawing, and through most of history since, drawing has been regarded as a means of pragmatic communication. The art historian Peter Cannon-Brookes put it that

> the depiction of such subjects [history, literature] was felt [in eighteenth and nineteenth century England] to be the most important purpose of art; it was not just William Blake who felt that all art worthy of attention should have a moral purpose.[2]

As humans became civilized, practiced agriculture, built cities, and engaged in trade, there developed a need for keeping records, not at first for such purposes as preserving the proceedings of the tribal council—that came later—but simply for commercial records. Who gave what, for which later payment was expected? Ambiguity is tolerated in art, even sought after, but not in commercial documents. A receipt for three donkeys must make it perfectly clear how many donkeys were transferred, at what price, and under what terms. This is why writing was needed. This is also an excellent example of what writing does for us. It enables communication over time. The invoice or receipt is intended to be read at a later time than it was written. A spoken record simply would not do the job.

Pictures and Pictorial Writing

The earliest known images created by a human being have been dated as early as 45,000 years ago. Since the ability to draw was available that early in our history, we must assume there was no need for writing until a relatively advanced commerce came into being that required it. Or could it be that for thousands of years no one got the idea of how to use artistic skill for business record keeping? This second alternative seems unlikely. Even without written records, people might give goods one day and expect compensation another day, and both parties had to have a reliable way of remembering how much was owed. Apparently, hunter-gatherers simply did not need such sophistication.[3]

As far as we know, the direct precursor of modern-day writing originated around 8000 BCE. There are disputes as to where, though Sumer, in what is now Iraq, remains the favorite. Some believe it may have been in Egypt, some in the Indus Valley. We will probably never know for sure. The reasons why Sumer is the favorite are that there is so much material from there available and there is a direct series of cultural links from Sumer to European and American culture.

The early forms of something like writing were tokens, in effect small sculptures formed from wet clay and then hardened. They took various geometric forms such as spheres or cubes. They were used to keep track of traded merchandise. If you lent two donkeys, you would expect your client to give you two donkey tokens as indications of a debt. The tokens for a transaction were kept in a hardened clay envelope.

Gradually, the tokens evolved into pictures of tokens, also impressed into clay that was then hardened. This was the beginning of pictographic writing, dated around 3000 BCE. The new writing provided a significant advantage over tokens by reducing the volume of material that had to be "filed." Creating the pictographic images would have been easier and faster than forming the three-dimensional forms. As the pictographic vocabulary expanded, scribes were able to portray abstract concepts as well as merchandise. Pictographic writing spread both to Egypt, becoming hieroglyphics, and to the Far East, becoming Chinese ideographic writing. In Sumer itself, pictographic writing evolved into cuneiform writing. Figure 3.2 shows some samples of the evolution of forms. Cuneiform was still pictographic but the images were made from combining several marks, all variants of a single symbol, easily made by a stylus in soft clay. This change would have made life much easier for scribes—fewer complex pictures to draw. It is estimated that the kind of transitions illustrated took some 2000 years to make.

Pictographic writing may appear primitive and restrictive to those who have learned only alphabetic writing. But Chinese society has shown no lack of intellectual inventiveness, technological development, or literacy. Still, the more images that were devised, the more the vocabulary increased, and the more there was to learn. Gradually, the need for yet another development was felt in the land where writing began.

Before we get to the alphabet we should acknowledge other forms of pictorial or graphic recording of information. One such is *wampum*. Another is the Incan *quipu*. Wampum was used by the Iroquois Nation as well as other Native American groups in present-day northeast U.S. and central and eastern Canada.[4] It consists (it's still to be found) largely of beads made from shells or wood and made up as strings or belts. Many of us were taught that wampum was a form of money and so it was for a time, but it was not only that. A wampum belt contains symbols that may be intended to carry a specific message, but the symbols do not have fixed meanings. To understand the message it is necessary to understand the story or the event the belt in a sense portrays.

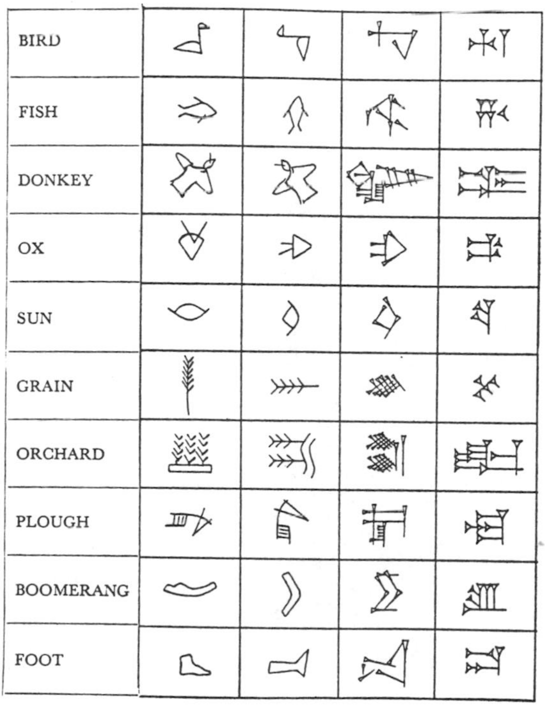

BIRD				
FISH				
DONKEY				
OX				
SUN				
GRAIN				
ORCHARD				
PLOUGH				
BOOMERANG				
FOOT				

Figure 3.2. *Evolution of cuneiform.* Shown here are some basic symbols and their original meanings that began as Sumerian pictographs and evolved into cuneiform writing. Although we can see the development, the cuneiform symbols bear little resemblance to the earlier pictographs. Reproduced courtesy of University of Chicago Press.

Some scholars do not accept wampum symbols as writing because they do not have established meanings. The belts might be used to commemorate a treaty or to establish the credentials of a messenger. Figure 3.3 shows an example called the Hiawatha Belt which has meaning but the meaning cannot be read from the belt out of context. It is necessary to know the accompanying story or context. The squares represent nations of the Iroquois Confederacy, from the bottom, they are the Mohawk, Oneida, Cayuga, and Seneca. The tree in the center represents the Onondaga Nation which had special status in the Confederacy. The lines connecting these five symbols represent the Path of Peace binding the nations together. (A sixth nation, the Tuscarora, joined the Confederacy later.)

Figure 3.3. *An Iroquois wampum belt*. This is called the Hiawatha Belt. The symbols have no fixed meaning. The viewer must know the context. Here, the squares represent four of the then five nations of the Iroquois Confederacy; the pine tree stands for the Onondaga Nation which played a central role in Iroquois ritual. The lines connecting these symbols represent the Path of Peace binding the nations. Photo courtesy Six Nations Indian Museum of Onchiota, N.Y.

The Incan *quipu* used complex patterns of knotted cords for accounting purposes and possibly more elaborate records. They have not been completely decoded.[5] Both these cases, wampum and quipu, demonstrate how different societies solved the essential problem of providing the kinds of records they needed.

Alphabet

As the societies of the Middle East grew more complex and technological they needed more sophisticated systems for recoding information. The alphabet simplified earlier forms of writing and had a huge impact on the societies that adopted it.

History[6]

The invention of the alphabet is generally attributed to the Akkadians, known to the Greeks and us moderns as Phoenicians, a seafaring Semitic people who lived in what is now Lebanon, near Sumer. Early alphabetic writing has been found also in modern day Israel and Egypt[7] but it was the Phoenician form that spread east and west, becoming the alphabets of Greece, Rome, the Arabic and Hebrew worlds, and India. The origin is generally considered to have occurred somewhere around 2000 BCE.

Sumerian culture was quite well developed long before the alphabet appeared. Sumerians were a technological people. As noted earlier, they had irrigation, the arch, contracts, some mathematics, the wheel, and some say also the sailboat. They invented the idea of recording history. Sumerians and the inventors of the alphabet lived close to each other and exchanged aspects of their respective cultures.

An alphabetic symbol represents a sound, not an object or idea. It must have been quite an intellectual feat on the part of whoever got the idea to listen to what a word sounded like, then try to represent the sounds as pictorial elements. But, of course, this was not an instant invention. It would have developed gradually over time. By way of example, classical Hebrew and Arabic did not have characters for all their vowel sounds. In some cases, the reader had to infer what vowels were intended and a string of consonants might have different meaning, depending on what vowels were assumed. There are elements of these languages that resemble the African drum languages in the high degree of context dependency. Eventually small marks were placed over or under a letter to indicate vowels in both Hebrew and Arabic. Vowel symbols as we know them in western languages are a Greek invention.

Why did the Phoenicians bother? Well, as pictorial and ideographic languages mature they acquire a great many symbols and this requires a great deal of learning by scribes. So, the alphabet is an early form of information technology, a new way to record information. In a way, it was analogous to cuneiform in that both methods use images representing full words made up from basic symbols. Both replaced earlier image forms, making the scribe's job easier. The alphabet went one step further than cuneiform. It was more efficient in terms of learning. If a person knew how to say a word, he could represent it in script without inventing new symbols. If someone could read, he could reproduce the sound of a new word without knowing its meaning.

The Impact of the Alphabet

Harold Innis, Marshall McLuhan, and Robert Logan[8] attribute the development of western science and its rational thinking to the development and use of the alphabet. The alphabet regularized and linearized writing. Symbols followed each other in proper order. Thoughts were laid out in the same way—in sequence. From this mode of thought came western science and technology.

We will never know for certain how we might have evolved culturally had we stayed with pictographic writing. The Chinese did but Logan points out that, for all their great accomplishments in art, technology, and administration, they did not develop basic science. He feels that basic science demands abstract thinking and universal rules and that the tendency to think that way is at least partially attributable to alphabetic writing.

The question of the alphabet's impact on western society is a complex one. I say "western" because so much of Asian language is not alphabetic. But we must remember that Europeans, including

European immigrants into the Americas and their descendants, almost all speak an Indo-European language the alphabets of which descend from early Semitic alphabets. So, the western world here means the Middle East, Europe, the Indian subcontinent, and the Americas.

The question raised by many scholars is whether the alphabet is responsible for generating the characteristics of abstract thinking (alphabetic characters are abstract; they represent only sounds, not meanings) and linear thinking (one concept or thought leads directly to another) that are so common in western cultures. From these ways of thinking came science and religious and ethical concepts common to most of the geographic area under consideration. Principal among these were monotheism and the rule of law. Also, as these modes of thought took hold, women's place in society and the pantheons of the various nations began to decline, to be replaced by male dominance on earth and masculine dominance among the deities in the heavens.

Leonard Shlain[9] points out that all this began with the alphabet. He also points out that males tend to be more left-brain oriented, more scientific, more logical, more one-thing-at-a-time. Females are more likely to be right brain dominant, more likely to see the "big picture," for example to see a scene, not a collection of individual things in it. And they tend to be concerned with relationships more than men are.

Could the alphabet have caused all this differentiation? Possibly. The alphabet would have reinforced male tendencies to be left-brained. Or, could a left-brain dominant segment of human society have found this new idea of an alphabet fascinating, more so than women who preferred pictorial "explanations?" Shlain and Logan are recommended for those wishing to pursue the question further.

What is undeniable is that the alphabet came right after the great string of accomplishments by the Sumerians and those societies that became alphabetic thrived. But, of course, "thrived" is a word used to describe ourselves, the cultural descendants of the alphabet inventors.

The development of writing gave us the ability to communicate over time. Not only could commercial records be kept from day to day or planting season to planting season, but some records remained in place from the time of their creation to this very day, giving us a sense of what life was like then. As writing matured, of course, writers began to write with the intention to communicate to unseen, unknown, future readers.

Other Languages

The writing we have been describing was used for recording natural language. Humans also invented written forms for expressing complex mathematical concepts and for music. Logan identifies five modes of language: speech (oral), writing, mathematics, science, and computing.[10] The first two are generally agreed upon; the last three might be thought of as *uses* of language, but it is interesting that Logan thinks of them as distinct *forms* of language. Mathematical concepts, for example, could be expressed in words, but a concise and precise notation vastly improves a mathematician's ability to grasp the concepts being presented and to continue to develop new symbols for new concepts. The language is almost totally meaningless to untrained readers, but a highly efficient means of communicating information among the initiated.

It is possible to express the symbols shown in figure 3.4 in words. The equation describes the normal frequency distribution, or the well-known bell curve that occurs so often in statistics. But it is more difficult to work with the concepts when expressed only as words. The notations used by mathematicians not only save paper but help them to think in symbolic terms.

Musicians tend to be able to reproduce musical passages after a single hearing, but to get an entire orchestra to play the right notes, correctly timed and stressed, requires some common form of expression of what is expected. This is done by writing the notes and instructions as to key and tempo. Figure 3.5 shows a portion of a musical score annotated in this way.

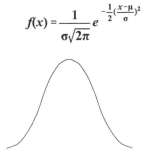

$$f(x) = \frac{1}{\sigma\sqrt{2\pi}} e^{-\frac{1}{2}(\frac{x-\mu}{\sigma})^2}$$

Figure 3.4. *Mathematical symbols.* This seemingly complex expression describes, to the knowing, the normal frequency distribution or bell curve, illustrated below, that occurs so often in statistics. For those who know the symbols it is far easier to read and understand than a verbal description of the relationships portrayed.

Amazing grace

Figure 3.5. *Musical symbols.* This is a portion of the music for the well-known hymn *Amazing Grace*. It shows the conventional musical notes, the key and tempo, and annotations telling what guitar chords should be played. Reproduced courtesy Wood Lake Books.

Science communication, in general, has its own meanings, procedures, and allowed and disallowed usages. This language, or group of languages, has its own vocabulary and syntax, although not generally as highly structured as mathematics. "Can notation really matter that much?" asks the physicist Hans Christian von Baeyer. He answers, "Of course it can. The way in which the building blocks of thought are designated profoundly affects the development of [a] discipline."[11] He offers as one example the immense benefit in all mathematical notation of the use of Arabic rather than Roman numerals. Arabic numerals actually were invented in India and were brought to the west by Arabs, as was a good bit of mathematics.[12] Curiously, they are used today in some but not all Arabic-speaking countries.

Chemistry and physics use notations somewhat like those of mathematics. The basic method uses one- or two-letter codes for chemical elements, combinations of element codes to describe molecules, and shows the effect of interactions among these (figure 3.6). Again, without such linguistic rules for making these complex symbols, communication among scientists would be far more difficult.

$$CH_4 + O_2 \longrightarrow CO_2 + H_2O$$

Figure 3.6 *Chemical symbols*. Chemists, like mathematicians, use a highly specialized notation to describe complex phenomena. This equation shows what happens when methane (CH_4), a compound of carbon and hydrogen, burns in the presence of oxygen (O_2). The result is carbon dioxide (CO_2) and water (H_2O).

In terms of history, computing is new on the scene, having for all practical purposes begun as a science and practical tool for solving complex problems only since the middle of the twentieth century. It can be thought of as either a branch of mathematics or of science in general. But it is also becoming so ubiquitous in terms of who uses it, and broad in terms of the nature of the uses, that it perhaps is deserving of its own place in the family of languages. Many people even with little formal training can talk of RAM, modem speeds, downloading, and rebooting. These terms are largely meaningless outside the computing field. The Internet and the World Wide Web are seen by some to be as radical a change in how we communicate among ourselves as was the introduction of printing around 1454. We are coming closer and closer to being able to communicate with computers in natural language, but usually the computer is far from having a human ability to understand and it is still necessary for users to consider this when "speaking" to a machine.

Recording Media[13]

The Nile River provides the life blood of Egypt, even today, by watering its crops. Among the plants that grow in its waters is, in modern terms, *Cyperus papyrus*, or papyrus. This is a grass-like plant with a pulpy stem. The pulp can be flattened, cut into strips, dried, and assembled into sheets, providing a convenient writing surface. The same material has also been used to make sails and cloth. Its earliest use for writing dates to the First Dynasty in Egypt (third century BCE). Papyrus is much lighter and thinner than hardened clay, making storage and transportation easier, but it lacks the durability of clay. It lasted as an important writing medium until supplanted by paper around the eighth or ninth century of modern times. Writing on clay was fault-tolerant in a way. Scribes could easily correct mistakes, at least while the clay was still wet. Writing on papyrus and later paper was not so easily modified.

As early as the second century BCE, parchment came into use for writing. This comes from the skin of an animal, generally sheep, goats, or calves, that has been cleaned, stretched, and scraped to make a smooth writing surface. It is believed to have originated in Pergamum, in what is now Turkey. Parchment is superior to paper and papyrus in lasting power, but its sheets are thicker and it was, and still is, much more expensive to produce.

The great challenge to papyrus came from paper—a word obviously derived from papyrus. It was invented in China in the second century CE. By the eighth century, knowledge of how to make it had drifted as far west as Baghdad and then gradually into Europe. During most of the period of the second through eleventh centuries Europe was hardly showing a great demand for written materials, but the Islamic world thrived intellectually in the latter part of this interval. Paper mills were found in Europe in the twelfth century before the invention of Gutenberg's movable-type printing press. These two, paper and the press, came around the beginning of the Renaissance and Reformation in Europe. The social changes and the technology that enabled the new ideas to be spread accelerated each other. A demand for printing in large quantities was created as was a new industry that could meet that demand. The press and paper were inventions well suited to their times and led to rapid growth in printing.

Paper is relatively simple to make in its basic form. It consists mainly of plant fibers, originally obtained by reducing cotton or linen rags to a pulp, then mixing with water. The result is poured onto a fine mesh screen through which the water can drip but not the pulp, and what remains is paper. It must of course be dried, compressed, and cut. The quality varies with the kinds of fibers used and the manner of compressing or calendering the sheets. The use of wood pulp came only in the nineteenth century and other fibers can be used as well. Paper, papyrus, and parchment are all resistant to changing what has been written. Parchment and some grades of paper would allow inked words to be scraped off the surface with a knife, but otherwise changes were made by crossing out, insertion, or recopying an entire page. This limitation on erasing meant that authors were well advised to think carefully before writing. Today, we have erasers (abrasives that scrape off the ink), various chemicals to cover up the mistake, bits of paper to paste over one, or a word processor to change the text and re-print it.

Paper is cheap, lightweight, low in bulk per page. Its lasting power depends on how it is manufactured, very old papers often outlasting newer forms because we have cut corners in the manu-facturing process, introducing acids which eventually destroy the paper that contains them. Modern newspapers, for example, virtually self-destruct in a few years. In terms of economics and convenience, paper has done so well that many people today cannot or will not consider that it might some day be replaced as it replaced other media. On the other hand, some modern papers may have high rag content or may have various coatings that resist wear and take some inks, particularly colored ones, better.

Paper is so versatile that it was hundreds of years of new technological development before anything newer came along.

The very latest developments in recoding media are related to computers. Basically they are of two types, magnetic and thermoplastic. Magnetic media include the familiar "floppy" disks, which were originally so flexible they seemed to flop over if held by one edge, and "hard" disks which are normally internal to the computer. Today's floppies are now as rigid as the hard disks. Magnetic tape is no longer common but had its day as a primary computer storage medium. All the magnetic media are basically alike. A coating of an iron-based material enables the surface to be magnetized. Information is stored in the form of tiny magnetized spots which roughly play the role of Morse code dots and dashes.

The second type of computer medium is the compact disk, a plastic material in which tiny deformations, too small to be seen by the human eye, are made by a laser beam essentially melting a spot that can later be sensed by another laser beam. These media have storage capacities that are staggering in comparison with any previous form, including microfilm. A CD-ROM (compact disk, read-only memory) can store about six hundred million characters. A 100,000-word book needs about one million characters, hence the disk could hold hundreds of such books. (There is always some "overhead," so probably not six hundred books, but maybe two to three hundred.) Newer DVD (*digital video disk* or sometimes *digital versatile disk*) can go to at least twice the CD capacity. This can mean a complete personal library can be stored on a single disk or all the assigned readings for a four-year university program.

Is there a catch? Well, yes. You cannot write directly onto these media nor read them directly. There must be a computer or other machine involved. One of the most frequently heard reasons for not publishing books in this form is, "Who would want to curl up by the fire, or in bed, with a computer?" But disks have proved very successful for storage and searching of encyclopedias and other reference works. There are those (including this author) who predict the eventual replacement of paper for books, magazines, invoices, bank checks, and the like by computer-related media. It will happen some day, just as we switched from clay to papyrus. But there is need for much improvement of the technology of reading and writing through a computer. Paper will stay for quite a while yet. Remember, though, that the essence of a book is not its paper but its content.

One of the possible replacements for paper, at least in some applications, is *electronic paper*, a form of which has been developed by Xerox Corporation.[14] Basically, it is two thin sheets of rubber in which are embedded tiny beads made with one color on one side, another color on the opposite side. An electric current can cause the beads to rotate, much as a changing electric current causes a tiny magnet to change the orientation of its poles. Thus, each bead becomes a pixel. Early versions had only two colors, but work is underway to make the product truly multicolored. Exactly how this might be used is still unclear but it does offer the possibility of giving readers a printed paper-like image that can be written over at high speed; therefore only one page is needed to read a book—a touch of a button could cause the contents of the next page to replace the current one. A book might come to consist physically of a small computer, say the size of a cell phone or palm-top, and a single sheet of "paper" which could be reprinted at the touch of a button, to simulate the turning of a page.

Another form of electronic paper is the use of a computer to display information without printing. We do this now, of course, but if the displays and the controls over searching, inserting, and page changing are improved, we may not need paper. This form of electronic publishing is being called *electronic books* or *e-books*.[15] At this stage, it is too early to see their effect on the world of books and communication. Basically, they are like a conventional desktop or laptop computer, only specialized for the purpose of display, downloading, annotation, and search within the text of the books. All these are desirable to many readers, especially those involved in research. Even with novels, we sometimes want to find a particular quotation or be reminded who some character is. But will this form replace the comfortable, familiar book? We cannot tell yet.

During all this time, from the ancient to the modern world, there have been media used for writing and drawing that were not permanent and not intended to be. Primarily used in teaching, they include sand (drawing in sand as Euclid did), slates to be written on with chalk or a similar substance, and today's "white boards," which are plastic or enamel written on with erasable dyes. The point is that for all our advances in permanent media, sometimes we just want to make some images for the moment, then erase them and make new ones. Saving them all would serve no purpose.

The Tools of Writing[16]

Writing requires a medium on which to record but also tools with which to create the symbols to be recorded. At the beginning, the tools were sharpened sticks or rocks. Later, they became mechanical, with the printing press and other writing machines. Today, still mechanical, we have the computer and its many ways of assisting in writing.

Hand Tools

The earliest tools would have been colored stones or sharp rocks with which to draw pictures on cave walls. Many of us, as children, would have found a bit of colored stone and used it to write or draw on such surfaces as a concrete sidewalk or macadam road. Not as good as oil paints, perhaps, but quite good enough for five-year-old artists as it must have been for the cave dweller.

Then we skip ahead to writing on clay. All that would have been needed was a pointed stick, not unlike a modern day wooden pencil. But as writing evolved into cuneiform, a more specialized tool was needed to make the triangle-with-a-tail marks that made up the cuneiform character. For that, it would have required a flat stick, something like a modern popsicle stick. Poke it into the clay, say vertically, and you have the beginning (figure 3.7) Turn it ninety degrees as you pull it to the right and the triangle is formed. Then drag it farther to the right and the tail is made. With a little practice this could be done quickly and oriented in any direction.

Figure 3.7. *Form of a Sumerian cuneiform stylus.* This is what an early stylus would have looked like. It was a hollow reed, cut a bit more than half way through at one end. Rather similar to a quill pen of a later age.

Variations on the stylus were the brush, still used in making Chinese characters, and the pen. The latter was once a feather whose shaft could be sharpened to a point and is hollow, allowing it to store a small amount of ink. This, in turn, evolved into the steel-nibbed pen, then the fountain pen, then the variety of fountain pens, ballpoints, fiber points, and magic markers we have today. The last of these is not much different from a brush with its own ink supply. The pencil, a form of pen, is a wooden shaft hollowed out to hold a thin stick of graphite. Earlier forms, dating back to Roman times, used lead to make the marks. When graphite was first discovered, it was assumed to be a form of lead, and the name *lead pencil* survives to this day.

The Printing Press[17]

Johann Gutenberg did not invent the printing press. Nor was he the first to use movable type but he brought the movable-type printing press into existence in the western world.

Printing, of a sort, was known in China as far back as the seventh century CE. This was printing from wood blocks into which reverse images of written ideographs were carved. The method is still sometimes used to print patterns on cloth. It is something like using a large rubber stamp. An ingenious people, the Chinese may not have been motivated to develop individual types for each character because there were so many. A modern printer using alphabetic type (or perhaps one of a hundred years ago, when typesetting was routinely done by hand) could stand before a case of type and easily reach every character. A Chinese printer would have been faced with tens or even hundreds of thousands of types.

The technology of paper making has been traced from China to the Middle East and then to Europe. A form of movable type was developed in Korea in the late fourteenth century CE, but there is no indication that Gutenberg did not conceive of movable type on his own. He was born around 1400 (some uncertainty here). Presses were already in use but only for printing graphics and as a part of the papermaking process. He was a goldsmith who would have known how to make precise objects out of metal. The plan was to design and make a piece of type for each character to be printed. These had to be of the same height, to a fine tolerance. He had to design the case in which to place the assembled type to be printed. He had to concoct ink that would adhere properly to metal, yet transfer easily to paper. This was different from the ink used with wood prints.

He was imaginative, dedicated, competent, but not rich. He had to borrow money and take on investors as partners. These transactions led to a number of lawsuits which deprived him of many of the financial fruits of his invention but provided a record of his activities. His first Bible was produced in 1454 (sometimes considered 1455, occasionally other dates). Thereafter, most production came from his partner-successors, Fust and Schöffer.

All this work was done during the early stages of the Renaissance and just before the Reformation in Europe. Learning was becoming more widespread, hence so was interest in reading. Printing spread rapidly in the fifteenth and sixteenth centuries; not as rapidly as today when we can see a book written and published in a matter of weeks after a significant event, such as the death of

Diana, Princess of Wales. She died on August 31, 1997. The Library of Congress shows forty-eight books about her and related to her death, published in 1997 or 1998. The point of this is that it is technology in publishing, telecommunication, and travel—all communication related—that enables books to be published so soon after an event. But the spread of books in Gutenberg's day was astonishing for the times. The world has never been the same since and probably never will be again.

Other Writing Machines

After the press, the next big step was the typewriter, patented in the United States in 1868. Mark Twain was an early adopter, becoming the first author to submit a typewritten book manuscript to a publisher. Eventually, the typewriter became a standard office machine and was often found in the home. A bit hard to learn, it produced printing of higher quality than most of us can produce by hand but not up to that of the printing press. It could make several copies at one time. A much used model is shown in figure 3.8.

Figure 3.8. *A typewriter of the early twentieth century.* This Underwood typewriter was built around 1915. Others similar to it began to be produced around 1900 and many were still in use until electric typewriters became popular in the 1950s.

The linotype could be considered a form of typewriter. Invented by Ottmar Merganthaler in 1884, it is a large machine. It does not fit on a desk but could consume a small room. The linotype does not print on paper. Instead it creates "slugs" of metal type, one line long, that go to make up the printing plates used on a large press. It is fed molten metal, which it forms into the selected type images as shown in figure 3.9. Later, the type is remelted and reused. This machine considerably speeds production of large printed works, such as the modern newspaper. The Sunday editions of the New York or Los Angeles *Times* would be impossible if type had to be set by hand, that is, by a compositor picking each individual piece of type from a case.

Figure 3.9 *The Linotype machine and its product.* This ungainly nineteenth-century machine was remarkably versatile. The operator typed the text on the keyboard on the viewer's left. As each line was completed, the type for that line was assembled, then cast into a *slug* made of lead. The slug was already justified and ready to be used. Several slugs are pictured at right. After use, the slugs were melted down and the metal reused. Photos courtesy Heidelberg Canada.

Computers and Writing

Not too surprisingly the next big step was the computer. The initial step in recording text in a computer is done essentially as typing and linotyping are done. Each character is represented by a key on a keyboard almost identical to that of a typewriter. Word processing computer programs give us what we have never had before, a means of typing text and easily and quickly correcting it. When the typist is ready a clean printed copy can be produced that shows no sign of correction. Newspapers are now published with stories that were never in paper form until the finished product came off the press—reporters type directly into a computer. Books are routinely written the same way. I typically write a rough first draft on paper, type it into my computer, and then edit again and again, before anyone else is allowed to see it. I have felt, in fact, that now that almost anyone can produce an elegant-looking page, readers of drafts are far less tolerant of small errors than they would be if given a handwritten or inexpertly typed version on yellow paper, clearly indicating a rough draft. Word processors gave us a rare example of a communications medium, the typewriter, being driven almost entirely out of use.

Is there an ultimate writing tool? The very latest is a computer system that can hear spoken words and render them as written text. Still imperfect, but they get better. Do these systems really understand what is being said? No. But court reporters, the champions of the stenographic world, tend to concentrate on hearing words and phrases, not on understanding what they mean and they reproduce text very well. Computers do the same, concentrate on the word not the meaning, and they may someday do as well as the human recorders.

Photography[18]

Photography is a means of reproducing an image, whether the image is of what we usually call a picture or of alphabetic characters, or print. It can be presented as *still images*, single images to be viewed independently, or as *motion pictures* in which a series of stills is presented to the viewer in such a way as to make it appear that objects in the image are moving. Recorded sound can be combined with motion pictures as it could with still images, but the latter combination has not been an important one, while motion pictures have, of course, been, highly important commercially, educationally, and artistically. The manner of presenting information through photography is of importance. Do we see the photograph in what has come to be called printed form, something we can hold in our lap or on the table top as we view it? Or, do we view it as projected upon a wall or screen as in a motion picture theater?

While we may think of photography as a relatively new technology, its roots go very far back in time. Consider Plato's famous parable about the men chained in a cave who could see only the shadows of other people. They could perceive the world outside a darkened cave only as two-dimensional, binary colored (black on cave wall) images. A different medium of conveying information led to a different perception of the world. Another example of how early the interest in something like photography developed is the experimentation with the effect of sunlight on silver nitrate in 750 CE. This did not lead to anything practical but it was continued experimenting along this line that eventually produced photography.

One of the earliest practical devices for displaying images was the *camera obscura*, a term meaning "dark chamber." It was a room or merely a box with a single small hole on one side through which light could enter. The light would be projected onto the far wall. The image from which the light came would appear sharply defined but upside down, as in figure 3.10. We do not know the originator or date, but these devices were used as early as the sixteenth century CE, initially as a means of studying an eclipse of the sun. The image of the sun projected onto the wall allowed it to be viewed without damage to the eyes. This method is still among the recommended procedures for watching a solar eclipse. Eventually users began to experiment with using a lens to project the image of a picture or a person onto the far wall, and this was the beginning of public showings of pictures, done now with slide shows or motion pictures.

The first of what we may think of as photography—recording an image made by light—was developed by Joseph Nicéphore Niépce in France in 1822. He called his device a *heliograph* and used it to deposit images on stone or metal (he was a lithographer) and eventually recorded an image of a scene on a pewter plate that took eight hours of exposure.

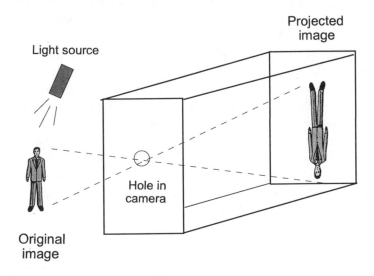

Figure 3.10. *The camera obscura.* Light from an object outside the camera enters through a small hole and is projected onto the opposite wall as a mirror image and upside down. The image can be quite sharp. Put a lens in the hole and some appropriately coated film on the far wall and you have the essence of a modern camera.

Today, we generally attribute photography to Louis Jacques Mande Daguerre, also of France. He produced the *daguerreotype* in 1839. This recorded images on a copper plate treated with silver iodide. When the plate was exposed to light and then to mercury vapor, the image would be developed. The resulting images were extremely sharp.

By the 1850s millions of daguerreotypes had been made. In 1849 a process of using a glass plate upon which to record the image was developed in England by Frederick Archer, but William Henry Fox Talbot, also British, who had invented an earlier process claimed Archer's invention was based on his own, the sort of claim we shall be seeing again and again. The glass plates made photographers more mobile, and Roger Fenton of England and Matthew Brady of the United States took their cameras to the field of battle—Fenton in the Crimean War and Brady in the U.S. Civil War. They made photography into a news reporting and artistic medium. Brady's photographs are emotionally moving to this day.

In 1871 Richard L. Maddox created a new form of chemical emulsion to be used with glass plates, which made it possible to take a picture "instantaneously." A year later Eadweard (or Edward, depending on source) Muybridge began to make multiple photographs of horses in action and was able to show that at some point a running horse does, indeed, have all four hooves off the ground. That may not sound important in itself but it began the use of photography in science to study phenomena that occur too fast for the human eye to observe accurately.

Microfilm is a photograph reduced in size by a factor of around ten or more. If that small, a page image can be hidden from police or customs officers, or attached to a pigeon. It has been used since the Franco-Prussian War of 1870-71. It is still often used to preserve documents, providing both a medium less prone to decay than many kinds of paper, and much smaller for storage.

In 1887 Thomas Edison devised a means of projecting a series of still images so as to appear to a viewer as if the pictured images were moving. He called it the *kinetoscope*. French photographer Antoine Lumière saw what Edison had done and brought the idea to his two sons, Auguste and Louis, who developed a combination

motion picture camera and projector in 1895. Then George Eastman began using transparent celluloid as the base for the photographic emulsion, making film easier to work with. Both Edison
and the Lumière brothers, Auguste and Louis, began to produce
and show publicly what came to be called cinema or "movies."
Eastman's celluloid film also made home photography popular.

In the twentieth century change was rapid, as with all other
communication technologies. We got motion pictures that included
sound, color, wide screens and even wider with IMAX that fills the
visual field of a viewer, stereophonic sound, and movies recorded
on videotape for home viewing. It is clear that from the Lumières
and Edison a great industry developed. Surprisingly, Louis
Lumière has been quoted as having said, "The cinema is an invention without a future,"[19] quite the opposite sort of thinking that
usually comes from inventors. For example, Thomas Edison said
about the same invention

> I believe that the motion picture is destined to revolutionize our
> educational system and that in a few years it will supplant largely,
> if not entirely, the use of textbooks.[20]

The Effect of Writing on Civilization

We have made much of the fact that writing began as a business
practice. It allowed the transactions of a business to be recorded as
an aid to management and as a means of resolving disputes over
past transactions.

Our modern world needs organizations, whether these be farms,
governments, or automobile distributorships. To run an organization requires knowledge of resources available, resources needed,
debts owed to others, debts owed to us, and schedules of important
events. From this basic information it becomes possible to deduce
what we will be able to accomplish or what is lacking and must be
found if we are to accomplish some task. Want to build a pyramid?

How much stone is needed? How much is available? How many workers are needed? How many are available? How long will it take to quarry and move the stones? Where will food be found to feed the workers? How much is needed? How and when will it be delivered? Mendelssohn[21] said that the workers were not slaves as that term was understood in the Americas, but that all citizens of Egypt owed the Pharaohs a certain amount of work each year for which they received food and royal favor. But the undertaking was seasonal. Life in ancient Egypt depended on the Nile River and its flooding to provide irrigation for crops. The work had to be scheduled around this reality. The point is that a great deal of management planning was needed. Imagine doing it without writing and systems of accounting.

Writing also allowed for the recording of laws. Among the earliest written codifications of law were those in Sumer around 2400 BCE, the Ten Commandments given to Moses, and the Babylonian King Hammurabi's Code popularly credited with being the first written set of civil laws. Babylon was in Sumer, which by Hammurabi's time was no longer so-called, but the tradition of having written laws had already been established by Hammurabi's time.[22]

Once laws are written they remain subject to differing interpretations. The legal and theological professions have long devoted themselves to debates over the interpretation of laws. In our British-American legal system, when any judge renders an opinion interpreting a law, it is recorded and is considered by other judges in any future interpretations of that law. Thus, writing has not made our laws fixed in meaning. They can be and are still being differently interpreted.

The future of writing is often debated. Some feel writing on a surface like paper is too deeply ingrained in our society ever to change. Others feel technology will change, possibly even doing away with alphabetic forms. It is really impossible to predict.[23]

Writing developed slowly from its origins as commercial tokens. As it evolved it became ever more important to society. The invention of the alphabet and much later of printing are of inestimable benefit to our society. Writing and printing became so important to the modern industrial world that ability to read and write are essential to all but the most elemental of jobs. The breadth of one's reading of the "best" of literature is often used as the measure of how cultured a person is considered to be, almost what you read is who you are.

Evolution was slow, as we might view it, beginning with three-dimensional clay images and becoming, for westerners, alphabetic and machine aided through printing presses, typewriters, and computers. For all the technical developments, the importance of writing to commerce and culture kept increasing. It may someday become primarily a means of displaying information on a computer or television screen.

Notes

1. General reading: Chappell and Bringhurst, *Short History*; Clapham, "Printing"; Diamond, "Blueprints and Borrowed Letters;" Eisenstein, *Printing*; Hooke, "Recording and Writing"; Innis, *Empire*; Logan, *Alphabet*; McMurtrie, *The Book*; Shlain, *Alphabet*; Ullman, *Ancient Writing*.
2. Cannon-Brookes, *Painted Word,* 7.
3. Marshack, "Art and Symbols"; Innis, "Media"; Schmandt-Besserat, "Earliest Precursor" (There are two articles with this title.); Yule, *Study*, 12.
4. Astle, "Wampum"; Tehanetorens, *Wampum Belts;* Warkentin, "In Search"; Williams, *Linking,* 50-53.
5. Asher and Asher, "Civilization without Writing."
6. Logan, *Alphabet*, 26-45; Yule, *Study*, 8-15; "Alphabet,"*Encyclopaedia Britannica*.
7. Maugh, "Was Kilroy There."
8. Innis, *Empire*, 57-81; McLuhan and Logan, "Alphabet;" Logan, "*Alphabet*."
9. Shlain, *Alphabet*.
10. Logan, *Fifth Language*, 63.

11. Quote is from von Baeyer, "Nota Bene." See also Ifrah, "*From One*," 428-489.
12. Ifrah, "*From One*," 428-489;
13. Blum, *On the Origin*; Innis, *Empire*, 113-169; "Media" and "Paper," *Encyclopaedia Britannica*.
14. See Guernsey, "Beyond Neon" for a brief description of the system; and Sheridon and Berkovitz, "The Gyricon" for a technical description.
15. Kahin and Varian, *Internet Publishing*; Meadow, *Ink into Bits*.
16. Adams and Butler, *Manufacturing*; Berry, "Printing"; Hoke, *Ingenious*; Kilgour, *Evolution*, 81-160; "Pencil," *Encyclopedia Americana*.
17. Eisenstein, *The Printing Press*; Kilgour, *Evolution*, 91; Printing Press, *Encyclopaedia Britannica*.
18. Burns, *The Complete History*; Hirsch, *Seizing the Light*.
19. Quoted in Burns, *The Complete History*, chap. 15.
20. Quoted in Soloway and Pryor, "Next Generation."
21. Mendelssohn, *Riddle*, 141-149.
22. Kramer, *Sumerians*, 83.
23. Crawford, Hurd, and Weller, *From Print to Electronic*; Logan, *Alphabet*, 227-247; Meadow, *Ink*, 229-238; Nunberg, *The Future of the Book*; Peek and Newby, *Scholarly Publishing*.

4

Visual Signaling[1]

Introduction

Nothing travels faster than light.[2] If you've been to a baseball game and sat in the center-field bleachers of a large stadium you may have experienced seeing a ball hit but not hearing the crack of the bat until the ball was flying over second base. Or, you must certainly have seen lightning then waited several seconds before hearing the roar of thunder that goes with it. These phenomena happen because light travels so much faster than sound. Light travels at 300,000,000 meters or 186,000 miles per second, sound at about 300 meters or 1000 feet per second. The actual speed depends on the medium through which the waves travel. It takes only a fraction of a second for light to travel from one point on earth to any other.

Primitive people seeing a fire on a distant hilltop, or someone waving, or a puff of smoke rising would think of the transit time as instantaneous, that is, it would seem to take no time at all. These three forms must have been the earliest forms of signaling over a significant distance. They predate recording, so we do not know when they were first used.

In Greek mythology Prometheus took fire from Zeus and gave it to humans, an offense for which he was severely punished. Fire became important to people in many ways: protecting them from cold, preparing food, driving away wild beasts, and signaling. It was used in communication because people could see fire or smoke from much farther away than they could see a person waving his arms like a semaphore. Of course, it took time to build a fire. If the signal were a matter of yes or no, the very existence of a fire or smoke would send the message. The elapsed time to send it would not be very great. If the intent were to send more complex signals, the sender had to light a fire, then to control the rising of smoke that made the symbols, and maneuver the burning material in a distinct manner. That could be rather slow.

Fire and Smoke

The use of fire as a signaling device is mentioned in some very old sources: an ancient Greek play, the Bible, the history of the Israelite captivity in Babylon, the story of Paul Revere, and George Custer's memoirs, among others.

Fire

In the first two examples below, the fire itself, or the fire being seen in a particular place constituted the signal. In Jeremiah 6:1, we read, "O ye children of Benjamin, gather yourselves to flee out of the midst of Jerusalem, and blow the trumpet in Tekoa, and set up a sign of fire in Beth-haccerem, for evil appeareth out of the north . . . ."

Aeschylus, the Greek playwright (525-456 BCE), wrote his play *Agamemnon*, about the Trojan War. We again hear of the use of simple fire signals, in this case to indicate the capture of Troy:

> I wait; to read the meaning of that beacon light,
> a blaze of fire to carry out of Troy the rumor

and outcry of its capture; . . .
(*A light shows in the distance.*)

Oh Hail, blaze of the darkness, harbinger of day's
shining, and of processionals and dance and choirs
of multitudes in Argos for this day of grace.
Ahoy!
I cry the news aloud to Agamemnon's queen,
that she may rise up from her bed of state with speed
to raise the rumor of gladness welcoming this beacon,
and singing rise, if truly the citadel of Ilium
has fallen, as the shining of this flame proclaims.[3]

The Talmud tells of more complex fire signals. After Israelites
were taken into Babylonian captivity, it became the custom for
those remaining in Jerusalem to signal those in Babylon to indicate
when a new month had started, which always marked a holiday.
The key was the appearance of a new moon and it was important
to catch the actual beginning of a new moon, not to be a day early
or late. The signal was a sequence of fires:

2. Before time they used to kindle flares, but after the evil doings
of the Samaritans they enacted that messengers should go forth.
3. After what fashion did they kindle the flares? They used to take
long cedar-wood sticks and rushes and oleaster-wood and flax-
tow; and a man bound these up with a rope and went up to the top
of the hill and set light to them; and he waved them to and fro and
up and down until he could see his fellow doing the like on the top
of the next hill. And so, too, on top on the third hill . . . until a man
could see the whole exile before him like a sea of fire.[4]

The chain of fires would have been about 850 kilometers long,
indicating some poetic license in the suggestions that all the fires
could be seen at once. Note that the coding system did not rely
simply on a fire existing or not existing, but on a pattern of movement by the sender and a similar pattern by a recipient to confirm
receipt of the signal. The Samaritans would at times intentionally
send a false signal. So, hanky-panky in transmission is not a new

thing. The situation was due to a schism between the Samaritans and Jews, the latter refusing to recognize the former as practicing Jews. Apparently in revenge for this treatment, the Samaritans intentionally sent signal fires on the wrong days. Note that the solution was to revert to an earlier and simpler transmission form—sending a messenger with the news.

Much later, Longfellow's epic description of Paul Revere's ride tells another tale of fire-based signaling:

> He said to his friend, "If the British march
> By land or sea from the town to-night,
> Hang a lantern aloft in the belfry arch
> Of the North Church tower as a signal light,—
> One if by land, and two if by sea;
> And I on the opposite shore will be,
> Ready to ride and spread the alarm
> Through every Middlesex village and farm,
> For the country folk to be up and to arm.
>
> . . .
>
> And lo! as he looks, on the belfry's height
> A glimmer, and then a gleam of light!
> He springs to the saddle, the bridle he turns,
> But lingers and gazes, till full on his sight
> A second lamp in the belfry burns.[5]

Here we have a classic communication system. The transmission medium is light, provided by fire, since the events being observed were expected to happen at night. There is a three-element code: one lantern if invasion is by land, two if by sea, and—critically important—no lanterns if no invasion.

Smoke

Smoke was used for communication by Native Americans living on the Great Plains in the center of the continent. It is a way to use fire during a bright day when flame might not have been seen easily from a long distance. The Great Plains are, of course, mostly plains but there are hills and often long vistas from the top of one to another. So it is natural that this method would be used by these people, but not by other native peoples who lived in heavily wooded areas of the east and far west where open lines of observation were not readily available.

General George Armstrong Custer came to a violent end at the hands of the Plains people in 1876 but he had time to record some observations of them. Here is his description of signaling by smoke:

> It is wonderful to what a state of perfection the Indian has carried this simple mode of telegraphing. Scattered over a great portion of the plains, from British America in the north almost to the Mexican border on the south, are to be found isolated hills, or, as they are usually termed, "buttes," which can be seen a distance of from twenty to more than fifty miles. These peaks are selected as the telegraphing stations. By varying the number of the columns of smoke different meanings are conveyed by the messages. The most simple as well as most easily varied mode, and resembling somewhat the ordinary alphabet employed in the magnetic telegraph, is arranged by building a small fire which is not allowed to blaze; then, by placing an armful of partially green grass or weeds over the fire, as if to smother it, a dense white smoke is created, which ordinarily will ascend in a continuous vertical column for hundreds of feet. This column of smoke is to the Indian mode of telegraphic what the current of electricity is to the system employed by the white man; the alphabet so far as it goes is almost identical, consisting as it does of long lines and short lines or dots. But how formed? is perhaps the query of the reader. By the simplest of methods. Having his current of smoke established, the Indian operator simply takes his blanket and by spreading it over the small pile of weeds or grass from which the column of smoke takes its source, and properly controlling the edges and corners of the blanket, the operator is enabled to cause a dense volume of smoke to rise, the length or shortness of which, as well as the number and frequency of the columns, he can regulate perfectly, simply by the proper use of the blanket. For the transmission of brief messages, previously determined upon, no more simple method could easily be adopted.[6]

There are several points of importance here. Elsewhere, Custer made clear his disdain for the Native Americans, whom he considered uncivilized. Yet, he grudgingly recognizes they had some admirable accomplishments and characteristics. He recognized that their transmission system was ingenious and well suited to the environment. The coding system met the most advanced standards of what Custer saw as the clearly superior "white man," although, as we have seen, similar systems were used by other peoples in other places to encode drum signals. A rather technical point is that the series of dot and dash equivalents were not an alphabet in the sense of the symbols representing sounds. The Plains peoples did not have phonetic symbols. The symbols were indirect, conveying no meaning in themselves. In more detail, here is how the signals were generated:

> First gathering an armful of dried grass and weeds, this was carried and placed upon the highest point of the peak, where, everything being in readiness, the match was applied close to the ground; but the blaze was no sooner well lighted and about to envelope the entire amount of grass collected, than Little Robe began smothering it with the unlighted portion. This accomplished, a slender column of grey smoke began to ascend in a perpendicular column. Little Robe now took his scarlet blanket from his shoulders, and with a graceful wave threw it so as to cover the smouldering grass, when assisted by Yellow Bear, he held the corners and sides so closely to the ground as to almost completely confine and cut off the column of smoke. Waiting but for a few moments, and until he saw the smoke beginning to escape from beneath, he suddenly threw the blanket aside, and a beautiful balloon shaped column puffed upward, like the white cloud of smoke which attends the discharge of a field piece. [*Note: field piece = cannon, artillery piece.*]
>
> Again casting the blanket on the pile of grass, the column was interrupted as before, and again in due time released, so that a succession of elongated egg-shaped puffs of smoke kept ascending toward the sky in the most regular manner. This beadlike column of smoke, considering the height from which it began to ascend, was visible from points on the level plain fifty miles distant.[7]

What is clear from this is that while the image of a smoke cloud traveled at the high speed of light, the generation of a puff of the correct duration, part of a sequence of puffs, was not a high-speed operation. But it was still faster and far safer than sending a messenger by horse. The plains are not burdened with a great deal of rain, but it is also clear that this is a fair weather, daylight activity. At night fire signals might have been used instead. In terms of actual signals or codes used, little information has survived. There seems not to have been a widespread common code, but Baldwin indicates that among the Jicarilla Apaches of New Mexico three or more puffs indicated danger. The more columns of smoke the more danger.[8]

Probably the best known modern use of smoke signals is in connection with the election of a pope. By tradition, whenever there is a secret vote of the College of Cardinals, the paper ballots are burned. The faithful, watching from outside the building, see black smoke come out the chimney if no one has received a majority vote. If a pope has been elected, the smoke is white. Straw is added to the paper ballots to make smoke, dry straw making white smoke, as it did for Little Robe, and wet straw making black smoke.

Flags, Semaphores, and Body Movements

Just as fire can be used to send the very simple message, "Hey, there's a fire here," colorful objects can be used in a similar way to send more information. Flags are used today to identify nations, military units, schools, and sometimes corporations. They were so used in the Middle Ages when identification of a knight in combat might have been a matter of life or death. A flag or escutcheon, a symbol used to identify a person or family, such as a coat of arms, does not actually extend visual range but it may present a sharp image that can be clearly seen and understood from a long distance. Figure 4.1 shows the escutcheon of ancestors of George

Washington. Residents of the District of Columbia will recognize the origin of the District's flag.

Figure 4.1. *The Washington family coat of arms.* This is the escutcheon of the family dating from before any of them emigrated to the American colonies.

Flags[9]

Flags are believed to have been first used in China, around 1100 BCE, where they served to identify the location of an important person, a king or other noble, in the field of battle. A flag is made of cloth and in battle or on parade is mounted on a staff. They began to be used in Europe during the Middle Ages. Going back somewhat, Roman legions carried symbols identifying the legion and its major subdivisions, just as do modern armies, at least on ceremonial occasions (figure 4.2). The Roman symbols were metal or wooden devices affixed to the top of a pole or staff, so it was not a great step to the cloth flag. The cloth displayed color vividly and was light in weight as well.

Early on, flags served to identify individuals of importance, as in China. Certain classes were permitted certain types of flags or none. For example, a major European noble might fly a long, tapered standard, with two points at its end. Certain knights could fly a square or rectangular standard. All knights could affix a pennon (a long, pointed flag) to their lances. And so on. The point was that size of the flag, its shape, its color, and the symbols depicted all had meaning. The meaning was primarily one of identity.

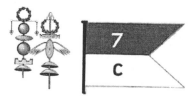

Figure 4.2. *Military unit symbols.* On the left are some Roman Legion insignia and on the right a cloth flag identifying a unit of the United States Army, C Troop, 7th Cavalry. Roman symbols @ Arttoday 2001.

As they evolved in Europe and the Middle East, national flags have tended to adhere to informal rules. They are almost always rectangular, with the horizontal dimension (the *fly*) longer than the vertical (the *hoist*). Two groups of colors are commonly used, one called colors, the other metals. The colors, again in Europe and the Middle East, are most commonly red, blue, black, green, and purple; the metals white (representing silver) and yellow (representing gold). Orange is rare, but it is seen in the flags of Ireland and India. An example of a rule amounting almost to syntax is that two metals should not appear side by side. The flag of the Vatican is a rare exception to this, being white and yellow in its background field.

Great Britain and the Scandinavian countries use a depiction of a cross. Some Islamic countries, Canada, Lebanon, and Israel use a religious, ethnic, or regional symbol on their flags, respectively the crescent, maple leaf, cedar tree, and star or shield of David. Figure 4.3 shows some modern national flags, all of which consist of three color bars. The flags of Canada and Mexico add a national symbol in the center. Countries formerly associated with Great Britain tend to have retained the red, white, and blue colors or a subset (Australia, New Zealand, and the United States use three colors; Canada uses only red and white). Islamic countries generally use red, green, black, and white, or a subset. Many African countries use red, yellow, green, and perhaps black. A solid black field indicates a pirate ship. A solid yellow flag warns of infectious disease on board a ship.

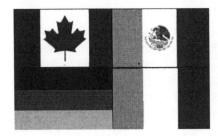

Figure 4.3. *Some national flags.* A common pattern for national flags is the tricolor—three horizontal or vertical bars. Another is to display a national or religious symbol. Shown here are the Canadian (top l.) and Mexican (top r.) flags with the three vertical color bars and a national symbol in the middle, and the German and Irish flags (bot., l. to r.) with color bars only. Change the colors or remove the symbol and a different nation may be represented.

The Netherlands flag consists of three horizontal stripes of red, white, and blue. When the Dutch revolted against their Spanish conquerors, starting in 1566, and eventually achieved independence, the tricolor became popular in Europe as a symbol of independence and republican rule. With variation in color and direction of the stripes, the tricolor was adopted by a number of countries including France, Belgium, Italy, Germany, and Hungary.

The custom of the flag indicating presence of a dignitary survives to this day. On a U.S. naval ship (and this has carried over to shore stations) when the captain or an admiral is actually present, his or her flag is flown. When that person leaves the ship, even if only temporarily, the flag is taken down. Not every officer rates a flag, but in the Navy, all admirals do and they are referred to, collectively, as flag officers.

In battle, because the flag was originally flown near the commander, its loss more or less indicated loss of the battle or the whole war. In a time when combat was conducted at close quarters, it was important to know who was who or at least who was on whose side. Today, this tradition survives in the requirements that soldiers wear uniforms and rank insignia for identification.

By varying how a flag is displayed, extra information can be conveyed. A ship surrendering in battle would take down its flag—strike its colors. A ship in distress might fly its flag upside down to attract help. To honor a dead person, a flag is raised to the top of its staff, then lowered down halfway. Ships also use flags of different colors and shapes to indicate individual letters or numbers and sometimes whole words or phrases.

The composition and use of coats of arms is subject to even more rules for how the symbol is formed and who may use it.

To summarize, flags are not a means of transmission, but they are a medium. Their shape, color, forms used therein, and even placement on the flagstaff can all convey meaning. The principal messages conveyed have to do with identifying who or what that flag represents: a rear admiral (two white stars on a solid blue field), the United Kingdom, a ship connected to deep-sea divers (red flag with a diagonal white stripe), mourning (half-staff flag), and so on. Clear, bright colors enable flags to be identified from a long distance. They are, of course, a daytime signaling device.

Semaphores

Semaphore comes from the Greek words for sign or signals (*sēma*) and for carry (*phorein*). It conveys the notion of carrying a signal or message. Today, we mostly use the term for a means of signaling with two colored flags held in the hands and moved to various positions to indicate letters and numbers (figure 4.4). The semaphore flags remain a useful way to communicate between ships at sea that are not too far apart and may lack radios or which must maintain radio silence.

Railroads once used a form of semaphore to communicate from the ground to the engineer of a moving train.[10] The railroad sema-

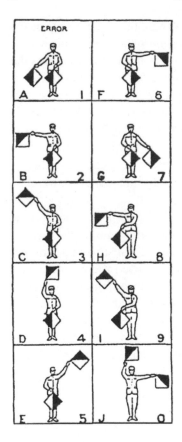

Figure 4.4. *Semaphore flags*. The color or pattern of these flags is not too important, except to help them to be seen. The position of the arms holding the flags is what sends the message. From *U.S. Army Signal Book*, 1912.

phore was a mechanical arm or perhaps one of several arms attached to the same post. They could vary in shape and color. By moving the arms to different positions, a signal could be sent to the engineer of a fast-moving train. Today, these have been replaced by colored electric lights, less susceptible to jamming by ice and more readily visible in the dark. Figure 4.5 shows an old railroad semaphore.

In both the old and new forms of railroad signaling, all that is required is some objects that can clearly be seen and an agreed upon set of meanings for various configurations of the objects—just as with smoke signals or national or heraldic flags.

Figure 4.5. *A railroad semaphore.* These, no longer used, consisted on a movable arm attached to a pole. The arm could be moved to any of two or three positions and the position as well as the color and shape sent signals to the engineer of a locomotive. Today's semaphores use electric lights. Note the telephone pole alongside the tracks, a common occurrence. Photo by C. Meadow.

One of the most successful pre-electric semaphore systems was developed by the Chappé brothers, in France in 1791.[11] They erected a device consisting of a horizontal bar atop a post or tower. A rope was used to tilt the bar one way or the other. At each end were two smaller arms mounted vertically on the main bar and able to be moved so the angle with the main bar was variable. The various positions of the bar and arms represented codes for letters. Figure 4.6 shows part of the code used. Like the Israelite signal fires of old, a series of these semaphores was erected on the tops of hills about five to ten kilometers apart, and messages were relayed from one to the next, then to the next, etc. Between Toulon and Paris, a straight-line distance of 700 km, there were 120 towers. A message could be sent from one city to the other in about 40 minutes, of course depending on the length of the message and the skill of the operators.

Visual signals sent by means of gestures or body movements have probably been used by human beings since before we had any records. A fairly obvious gesture is raising a club as a threat to bash in another's head, the meaning of which tends to be readily understood by anyone.

A system of mostly hand signals was widely used by the Native Americans whose smoke signals were described earlier. Alan R. Taylor[12] attributes its origin to people from the Gulf Coast area but

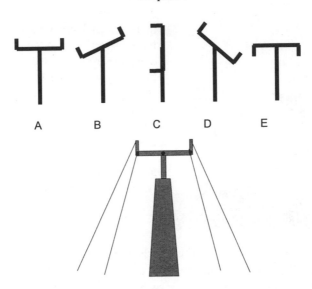

Figure 4.6. *The Chappé semaphore*. This apparatus was older than the railroad semaphore and based on the same principle. However, the intent was to send general messages, which required more arm positions to make for more codes. Shown are a few arm positions, the letters they represent, and a sketch of the kind of tower used, with the ropes needed to change position of the arms.

indicates it was widely used throughout the plains and by tribes with different spoken languages. Custer agreed with Taylor, pointing out that

> Almost every tribe possesses a [verbal] language peculiarly its own, and what seems remarkable is the fact that no matter how intimately two tribes may be associated with each other, they each preserve and employ their own language, and individuals of the one tribe rarely become versed in the spoken language of the other, all interconnection being carried on either by interpreters or in the universal sign language.[13]

All these methods of signaling are still in use to some extent: traffic signals on the road, red and green running lights on ships and airplanes, blue lights marking the edges of an airport runway. Fire has largely been displaced by electric lights, but these are nothing more than highly contained fire—a bit of metal or some gas heated to incandescence. McLuhan saw electric lights as communication media, not so much in themselves as for what they illuminated. "The instance of the electric light may prove illuminating in this connection. [*He loved puns. CTM*] The electric light is pure information. It is a medium without a message, as it were, unless it is used to spell out some verbal ad or name."[14] In defense of this thought, he offered the following from *Romeo and Juliet*:

> But soft! what light through yonder window breaks?
> It is the east, and Juliet is the sun. . . .
> She speaks, yet she says nothing.

New methods of semaphoring have been devised, such as sign language for the hearing impaired (figure 2.1), signs, lines and lights on highways to communicate silently with drivers. Sign languages for the hearing impaired are of two kinds. One is alphabetic and one pictographic. Those who know the pictographic form can communicate faster, but there is much more to learn.[15] These all amount to use of body movements as semaphores.

Even the computer world has adopted visual, nonverbal signaling using little figures called *icons*, from the Greek *eikon* (meaning image), shown in figure 4.7. They tend to take up less space on crowded computer screens than the words they represent. My own experience has been that these icons are not usually direct, although their designers may have intended them to be. I find myself learning and memorizing that a particular array of color elements has a certain meaning. I cannot count on being able to interpret the meaning of one new to me.

Visual signaling must be as old as the other basic human mode, making noise. It is still in use and is receiving a great deal of attention as we have, for example, begun to pay formal attention to body language, still use hand signals for directing traffic and orchestras, and use colored lights to direct cars, trains, and airplanes on the ground. It is a good example of the notion that communications media rarely go completely out of style, although they may change as new ones are developed.

Figure 4.7. *Modern computer icons*. These symbols are used with modern computers to designate functions or programs that can be invoked by the user. The alternative would be to display a box with the name of the program but most computer users seem to prefer the pictorial images. Photo by C. Meadow.

Notes

1   Other writers do not use this classification of information about communication, so there are no general references. Readers looking for more detail should consult the individual citations to follow.

2. Most of us will have been taught that the speed of light was absolute: nothing in the material universe could travel faster. Recently, some

physicists have begun to question that. The implications to communication of a means of transmission that is faster than light, and not dependent on wires or the like, is hard to contemplate. Johnson ("Quantum Feat") reports on an atom in two places at one time, Gibbs ("Faster") and Schatzer ("Speed") question whether travel faster than the speed of light is possible, in rather technical articles.

3. Aeschylus, *Agamemnon*.
4. *Mishnah*, 189.
5. Longfellow, "Tales," 606-609.
6. Custer, *Life,* 226.
7. Custer, *Life,* 217.
8. Baldwin, *Talking Drums*, 62.
9. "Flags," *Encyclopaedia Britannica;* "Heraldry," *Encyclopaedia Britannica.*
10. "History and Development of Railway Signaling."
11. Shaffner, "History."
12. Taylor, "Nonverbal," 349.
13. Custer, *Life*, 15. This language is described in some detail by Taylor ("Nonverbal") whose article is readily available and others that will be harder to find, e.g., Clark, *Indian Sign Language* and Mallery, *Collection*.
14. McLuhan, *Understanding*, 8.
15. Costello, *Signing*.

5

Transportation as Communication–1[1]

Introduction

Let us go back again to the beginning of civilization and then just a little forward to the time when people began to feel the need to communicate with others in distant villages. This could have been for trade, ritual observances, or assistance in war or hunting. The first option would have been to send someone to carry a message, as is still done. But what if there was a river to cross, an expanse of ocean, a desert, or a mountain range? It would not have taken an engineering genius to sit on the trunk of a fallen tree in the river and propel himself using his hands or a pole or wide piece of wood. That might have solved the short-run river problem. For longer trips or travel over land, there was the need to get help from an animal or invent a mechanical contrivance more maneuverable and able to carry a bigger load than a log in the river.

Transportation played a large role in the development of communication because that was how people reached out to others far away. From quite primitive beginnings there developed the yearning for ever faster means of sending messages. If we consider the four-hundred-year period from about 1450 to 1850, roughly from the invention of the printing press to that of the telegraph, most progress in communication came from transportation. Arthur

C. Clarke, the man who first conceived of the communications satellite, said, "When Queen Victoria came to the throne in 1837, she had no swifter means of sending messages to the far parts of her empire than had Julius Caesar—or, for that matter, Moses."[2] Railroad historian Thomas Clarke, writing in 1889, said, "The world of today differs from that of Napoleon Bonaparte more than his world differed from that of Julius Caesar, and this change has chiefly been made by railways."[3] They're both debatable, because both Victoria and Napoleon had ships that could cross the oceans but Caesar's ships could not venture so far. The telegraph, telephone, and elevator (enabling skyscrapers) had been invented when Thomas Clarke wrote, but had this been written forty years earlier, it would have been more accurate. Because the history of transportation is so large, we're going to skim through it rather quickly, concentrating on transportation media as carriers of messages rather than freight. In this chapter we cover transportation media before the ages of steam, electricity, and internal combustion engines. In the next chapter we cover developments based on these inanimate power sources.

Animals and Roads

The first step in training animals as carriers of communications or as messengers bearing communications was probably for some brave soul to jump on the back of a horse, camel, or donkey and eventually convince the animal to accept this burden and head in the right direction. The training of animals as beasts of burden contributed to the development of agriculture, which included using cultivated animals as sources of food. What we know about it suggests that this, like much else, first happened in the Middle East. The date was around 8000 BCE. Horses and camels were native to this area and even in the deserts there was enough forage to keep the animals alive.

Many years ago, the Freuhauf company, manufacturers of trailers for trucks carrying freight, advertised that, "A truck is like a horse. It can pull more than it can carry." The ad showed a horse with a huge load on its back and another one pulling a larger load on a wagon. This was known to early man, too. Our assumption is that the earliest form of vehicle would have been a pole or two poles, attached at the front to a horse or ox, with the back end left dragging on the ground behind the animal. A load would be placed across the poles. There would still have been lots of friction but the beast could, in fact, pull more than it could carry. Sleds have long been used for travel over snow and a sled-like, animal drawn vehicle was found in the vicinity of Sumer and dated to about 2500 BCE.[4] The version used by native North Americans was called a *travois* (figure 5.1), a word of French Canadian, not Indian, origin. For carrying messages, even this device was not necessary; the animal needed only to bear a single rider.

Figure 5.1. *A travois, freight and passenger carrier for unpaved roads and unwheeled vehicles.* Such a contrivance was used by North American natives to haul heavy loads in terrain without real roads. It was primitive but effective. Photo © 2001 www.arttoday.com.

One of humankind's great inventions was the wheel, believed to have originated in Sumer around 5000 BCE (around the same time those extraordinary people invented writing). It brought farther and faster travel. It also required roads or at least trails. The evolution and use of wheeled vehicles had to go in parallel with the evolution of roads, and the joint evolution of the two led to more travel and trade, hence more need for communication and transportation, hence more opportunity for trade. And so on. This cycle is still going on.

Pigeons

Pigeons were the first high-speed message delivery service. Called carrier or messenger pigeons, they were used in ancient Egypt, Greece, and Rome. In more modern times the following story about Nathan Rothschild is often heard, to this day:

> It is a well known fact in this country [England] that the London house of Rothschild used carrier pigeons in 1815 to obtain information of the course of the events on the continent, and thus was able to receive the news of the defeat of Napoleon at Waterloo three days before the English government did and to buy up largely English government stock at its then depressed price, and sell it at an enormous profit after the rise which took place when the news became generally known.[5]

Other historians have different versions, the main one being that a Rothschild agent on the Continent simply heard about the battle's outcome and carried the news on the next commercial packet boat to England.[6] Because the proponents of this latter version can name names, it carries more credibility. Whichever version is true, it vividly illustrates the value of information and its speedy delivery.

Paul Julius Reuter established a news service in 1849 that still bears his name. He used pigeons and the electric telegraph to achieve high-speed news delivery. As mentioned earlier, during the Franco-Prussian War, Paris was surrounded by the Prussians. The main way to get messages in and out was by air: balloons out and pigeons in. That may have been the high point in history for carrier pigeons. They were used in World War I but apparently not to any great effect. By that time, machines could fly.

Mail Service and Roads[7]

Mail and animal power were closely linked until the steam locomotive made it big in the mid-nineteenth century. The earliest known formal postal service was developed in Egypt around 2000 BCE and such a service was highly developed in Persia, under Cyrus (sixth century BCE). It was the historian Herodotus who characterized the workers with these now famous words, "Neither snow, nor heat, nor gloom of night stays these couriers from the swift completion of their appointed rounds."[8] Today this expression is found carved into the facade of the General Post Office in New York City.

The Roman Empire needed communication, much by ship, but also much overland. Roman road-building skills are well known. Good roads enabled fast, reliable animal-drawn vehicular travel. They built 50,000 miles of roads. The first and most famous, the Appian Way, was begun in 312 BCE. They established relay stations for exchanging tired animals. Messengers were able to cover as much as 170 miles in a day. As the empire declined, following the barbarian invasions, so did travel. This was one of the reasons for the darkness of the Dark Ages.

The thirteen British colonies which were to become the United States were mainly settled along the Atlantic coast. Roads, even between major cities, were minimal.[9] There was a report of a teacher in 1704 who had to travel from Boston to New York by horseback because the roads did not permit passage of a stagecoach. Benjamin Franklin, deputy postmaster general of the colonies, wrote to the British House of Commons that, "no posts went to the inland towns." Mail to areas away from the coasts had to be carried by friends of the sender who happened to be going in the same general direction and service was therefore highly irregular. Some inland mail was carried by friendly Indians. The frontiersman Daniel Boone led a party in 1775 that blazed the Cumberland Trail through the Appalachian Mountains.

After independence development accelerated. The need for communication among the newly formed states pushed the development of a system of roads first for stage coaches, then the motor car. Early in the nineteenth century the Cumberland Trail was made into a surfaced highway that carried mail as far as present-day West Virginia and served as the main route west for the newly expanding nation. Thereafter, highways grew at a great rate, culminating with the Interstate Highway System, begun in the 1950s.[10] In the early 1800s there were canals, then railroads, so that mail and trade were ever faster and safer.

The Pony Express[11]

One of the most romantic stories in the whole history of communication is that of the Pony Express. To put it in context, it began operation on April 3, 1860. At that time California was booming following the gold rush of 1849. The Civil War had not yet broken out but it was brewing. Both slave and free states wanted California in their camp. The eastern establishment needed communication with its western territories. Wagons took around three weeks to make the trip from Missouri, although there was rail and telegraph service available from the east coast to Missouri.

William H. Russell, Alexander Majors, and William B. Waddell, owners of a freight company, formed the Central Overland California and Pike's Peak Express Company to provide a service for carrying mail by horseback from St. Joseph, Missouri to Sacramento, California. Relay stations were set about 15-20 miles apart. At such a station a rider would change horses and continue until eventually replaced by a fresh horse and rider (figure 5.2). The inaugural trip west from St. Joe took just under 10 days, the fastest trip 7 days and 17 hours. The service was not unique in history. Genghis Khan set up a similar service with relay stations 25 miles apart and riders able to cover 300 miles in 24 hours.

There were risks. An advertisement for riders stated a preference for orphans, under eighteen and willing to risk death daily. Risks from what? The movies made it come mostly from Indians but "road agents" or hold-up men were by far the greater source of trouble. Because of these threats, the riders were paid well.

For how long did this marvelous service last? From all the publicity, one would think many years. Actually, it was eighteen months. The last ride was taken in October 1861. A great success? No, a financial disaster for its backers. Why so short a life? New technology—the telegraph. A telegraph line to the West Coast was completed from St. Joseph in that same month of 1861 and there was no longer a need for men and horses to carry the mail. The telegraph was far faster and easier on the orphan population.

Figure 5.2. *The Pony Express*. A rider arrives at a way station in order to change horses. This will be done as quickly as possible, the fresh horse ready to go as the rider approaches. This storied system of communication existed only for eighteen months. Detail from *Rocky Ridge Station* by William Henry Jackson, reproduced by courtesy of the Scotts Bluff National Monument, U.S. Department of the Interior.

Muscle-Powered and Sailing Ships

The earliest evidence of boats found in Egypt dated from the fourth millennium BCE. There is some evidence that the Sumerians also had this technology. Consider again the pyramids and the Nile. The massive building stones came from upstream. They could have been moved short distances over land by use of round logs placed ahead and the rocks rolled over them or by sliding the stones along a paved road. But boats did the long-distance haulage. The motive power was initially human rowers, although the Egyptians later developed sails.

The Romans developed oar-power to a fine art (humans pulling on oars; see figure 5.3). Remember Ben Hur being chained to an oar in the eponymous movie? But this mode would not serve well in Atlantic storms. Eventually, sail replaced rowing for all but the smallest vessels. And sail lasted a long time. The design of hulls and sails changed, but the basic concept lasted from almost prehistoric times to the nineteenth century CE and is still used today, although now mostly for pleasure craft. Figure 5.4 shows what Columbus's flagship was believed to be like. Such ships opened European trade with the Far East, found what became America, conducted trade across the Atlantic, and spread various European cultures all over the world.

Figure 5.3. *Motive power in a Roman trireme, first century CE.* An early form of a powered ocean vessel. Although the ship has a sail, it also has three banks of oars protruding from the sides of the ship. Motivation for the slaves who did the rowing came from a man with a whip. Photo © 2001 www.arttoday.com.

Figure 5.4. *The Santa Maria, fifteenth century.* This is a rendering of Columbus's flagship whose exact appearance is unknown. He was able to cross the Atlantic but lacking a measure for longitude, did not really know at first where he had landed. Photo © 2001 www.arttoday.com.

Sail, alone, did not conquer the oceans. It was also necessary to develop navigational instruments. Basically, a mariner had to know in which direction the ship was heading and where it was at a given time. The heading came from a magnetic compass. Location came in two parts, latitude and longitude. Latitude could be determined by measuring how high over the horizon the sun was at noon. Tables—early databases—were constructed to show the expected angle at various dates and latitudes. Sailors used a sextant, or similar instrument shown in figure 5.5, to measure the angle of the sun. Then, knowing the date, they would look in the table and say, "Aha, we must be at x degrees north (or south)." They did have to know which hemisphere they were in.

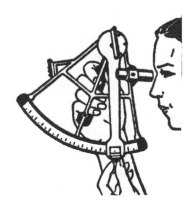

Figure 5.5. *Sextant.* This instrument measures the angle between the sun and the horizon. At any given time of day and latitude, we know from astronomical tables how high the sun will be. Similarly, knowing how high it is at a given time enables the navigator to tell what the latitude is. Photo © 2001 www.arttoday.com.

The sextant grew out of an ancient instrument called an *astrolabe* which was improved for use in navigation in the sixteenth century by Pedro Nunes, a Portuguese. It was in this period that the Portuguese led the European world in ocean exploration. Modern-day mariners have exactly the same information requirements, but quite considerably different instruments for determining them. See more on this in chapter 11.

Longitude, the angular distance east or west of some starting line, presents a different problem. Today, we use the meridian going through Greenwich, England, as the starting point. It could be any other. If we know what time it is in Greenwich when it is noon locally, we can know how many hours we are from Greenwich and can translate that into distance and distance into degrees of longitude. All this takes both the ability to measure latitude and an accurate, reliable clock. But making a clock that would stay accurate and reliable through the pounding of ocean waves on a small ship was no small problem when trans-Atlantic and Pacific travel began. There was a lot of ocean navigation before the clock problem was solved, but such a voyage carried a great risk of the ship getting lost. Once there was a solution, navigating over vast oceans lost much of its terror. The successful clock is attributed to John Harrison of England in the 1770s, produced in response to a competition established by the British Parliament.[12]

As the eighteenth century drew to a close, sailing ships and navigational devices were well developed. Steam engines had been developed but were not quite ready to be used to propel anything. Electricity was still a laboratory phenomenon. Roads were becoming well developed. Mail services were, at least to some extent, regular. All mail had to be transported until the telegraph came along to permit transmission instead of physical transporting.

After the start of the industrial revolution in the late eighteenth century and just before steam came fully into use early in the nineteenth, there was a flurry of canal building activity in the United States, Great Britain, and Europe. As a major transportation form, it was short-lived but continues to exist and to be used to this day. Canal boats were animal powered (figure 5.6). It would be hard to maneuver a sail boat in so narrow a channel, and steam-powered boats must take care not to destroy the banks by churning the water too much and this maintained the advantage of animal-powered boats. Today, the New York State Barge Canal, a rework of the Erie Canal of 1825, still functions, using motor-driven tug boats, but with relatively little commercial traffic.

Figure 5.6. *A canal boat, nineteenth century.* Motive power for nineteenth century canal boats was a mule or donkey. This one is on the Erie Canal in New York. Drawing by Lisa Mates, reproduced courtesy Erie Canal Museum, Syracuse, N.Y.

We are now ready to see a fabulous period in history, when steam and then internal combustion took over the powering of transportation vehicles on land and sea, and then in the air. Things went farther, faster, and cheaper. Just as those new developments must have seemed the ultimate to our dazzled grandparents or great-grandparents, along came electronics. But, we're getting ahead of ourselves.

Notes

1. Williams, *The History of Transportation*; "History of Transportation," *Encyclopaedia Britannica.*
2. Clarke, Arthur, *How the World Was One,* 3.
3. Clarke, Thomas C., "The Building," 1.
4. Cole, "Land Transport"; Forbes, "Roads to c.1900."
5. Cameron, *An Aid to National Defence,* 2.
6. Morton, *The Rothschilds,* 48-49.
7. Sidebottom, *The Overland Mail,* 131.
8. Cullinan, *The United States Postal Service,* 13.
9. Cullinan, *The United States Postal Service,* 14.
10. This system of highways was created by the National Defense Highway Act of 1956 and is sometimes referred to by the name of the bill, sometimes as the Interstate Highway System, and more recently as the Eisenhower Interstate Highway System. Mr. Eisenhower was president when the bill passed. Senator Albert Gore Sr. of Tennessee was a major sponsor.
11. Bradley, *Story*; Cullinan, *The United States Postal Service.*
12. Boorstin, *Discoverers*, 46-53. See also Umberto Eco's novel *The Island of the Day Before*, about perceptions of the longitude problem in the eighteenth century.

Part 3

Steam, Internal Combustion, and Electricity

Our story now enters the period when mechanical power in the form of steam and petroleum-based internal combustion engines drove most transportation systems and electricity enabled the sending of messages over thousands of kilometers without travel. When we think about inventions in communication, other than transportation, the electricity-based capability to transmit messages without physically carrying them, will rank with writing, the alphabet, and printing in the great technological developments of mankind.

We learned much about electricity in the nineteenth century, and, by the end of it, were using this new stuff to power factories, light homes and streets, and drive streetcars, as well as to carry messages by telegraph and telephone. People of that century must have been as dazzled by their age's developments in technology as we were in the twentieth century by automobiles, airplanes, radio, televisions, atomic energy, and computers.

6

Transportation as Communication–2[1]

Introduction

The persistent sail hung on for another 150 years after Harrison solved the longitude problem, finally to be largely replaced by steam and internal combustion. In the eighteenth and early nineteenth centuries a number of people experimented with steam engines for propelling railroad trains and ships. During the same period there was intensive study and experimentation with electricity. Electricity eventually served two roles in communication. It was first used for message transmission and later for powering mobile machinery such as motorcars and trains. The two forms of power, steam and electricity, profoundly affected society.

Steam Power

Thomas Newcomen (England, 1663-1729) and James Watt (Scotland, 1736-1819) are most frequently mentioned as originators of the steam engine in the eighteenth century. Actually, long before that time a Greek in Alexandria, Egypt, discovered that expanding steam could provide a motive force. His name was Hero. A prodigious mathematician and inventor, he developed an interesting but not particularly useful machine in the first century CE that used steam to turn a device with a nozzle on the end of a tube (figure 6.1). But, while interesting, it did not give enough power to be harnessed for practical use. Newcomen built his practical engine in 1706, but its invention is popularly attributed to Watt, who built a better one in 1765. These early engines were used for stationary purposes, such as pumping water out of mines or powering factory machines. As their size decreased and efficiency increased, they began to lure builders of railways and ships.

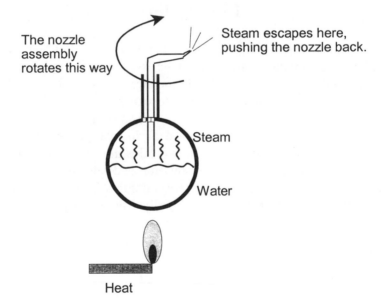

The nozzle assembly rotates this way

Steam escapes here, pushing the nozzle back.

Steam

Water

Heat

Figure 6.1. *Hero's "steam engine."* Water in a vessel is heated to boiling. Steam goes out the tube in the neck of the vessel at the end of which is a nozzle. Escaping steam pushes backward on the nozzle causing it to rotate. There was no way to make use of the mechanical energy so generated.

A steam-powered automobile was built in France in 1769. The word *chauffeur* derives from the French word for heat; the chauffeur was responsible for producing the heat for the car's engine. But automobiles, as we think of them, powered by an internal combustion engine, were yet to come. The invention of this mode of transportation was to become of huge importance in the development of technology and society.

What developed from steam and its later competitor, internal combustion, were vehicles that could cross the oceans or traverse large land expanses with previously unheard of speed and reliability. Long distance travel lost much of its senses of both menace and glamor. Travel across cities became common. Workers were no longer restricted to living within walking distance of their work.

Steamships[2]

Robert Fulton (United States 1765-1815) gets popular history's credit for the first *successful* steamboat, the *Clermont*, in 1807 (figure 6.2). The first steamboat appeared in France in 1783. By 1809 several countries were operating them. The first steam-powered vessel to cross the Atlantic was the American *Savannah* in 1819. But *Savannah* was a sailing vessel with auxiliary steam power. The first completely steam-powered crossing was made by the British ship *Sirius* in 1838.

Sailboats are now mainly pleasure craft. You can book passage on one to cruise the Caribbean or other waters, but rarely to cross an ocean. Figure 9.8 shows such a vessel, able to sail the ocean but now restricted to pleasure cruising. Modern ships driven by internal combustion, at the peak of their usage, could cross the Atlantic in three and a half days, the passengers (anyway some of them) basking in great luxury. Table 6.1 shows the rate at which technology cut the speed of a vessel crossing the Atlantic, from the days of Columbus's *Santa Maria* to the twentieth century's *S.S. United States,* holder of the transatlantic speed record and shown in figure 6.3. Speed is given in knots where one knot is one nautical mile per hour and a nautical mile is 2000 yards. The very latest technologies for powered boats are hovercraft and hydrofoils which can go at speeds ranging from 30 to 100 knots, but need more sheltered water than the oceans offer.

Figure 6.2. *The first successful steamship, 1807.* The *Clermont* was Robert Fulton's ship that made steam vessels practical. Photo © 2001 www. arttoday.com.

Figure 6.3. *The S.S. United States, undefeated transatlantic speed champion.* Built in 1951, the United States set the speed record at 35.6 knots but soon went out of service when this means of passenger transport went out of style as air transport improved. Photo courtesy S.S. United States Foundation.

Year	Ship (Country)	Speed (knots)
1492	*Santa Maria (Spain)*	1.8
1620	*Mayflower (GB)*	1.9
1819	*Savannah (US)*	4.6
1838	*Sirius (GB)*	9.0
1851	*Pacific (US)*	13.0
1907	*Mauritania (GB)*	27.0
1912	*Titanic (GB)*	21.0
1937	*Normandie (France)*	32.1
1952	*United States (US)*	35.6

Table 6.1. *Transatlantic crossing speeds*. These figures are derived from a ship's speed or from the time taken to cross the Atlantic. Actual times would vary much with weather, especially in the days of sailing ships. But it can be seen how much the speeds, hence crossing times, improved with improvements in motive power.

Railroads[3]

Railroads, by another name, were in use well before the steam engine. The Romans had railways, consisting of a path with tracks made of stone. In the fourteenth century, in what is now Slovakia, a railway was built of wood and used to draw wagons laden with ore from mines. They moved over a surface that offered less friction than a dirt or even a paved road. The motive power came from people. Even a donkey would have been an improvement. The first steam-powered railway locomotive was built in England as early as 1802 and one in the United States in 1805. But serious transportation by this mode did not begin in the two countries until 1825 and 1831, respectively.

Things developed slowly until around 1850 when almost explosive growth began, to be continued until well into the twentieth century. Figure 6.4 shows a construction crew at work, with a telegraph crew installing lines next to the tracks. In the United States, the Civil War helped, with its need to move troops and supplies. The Confederacy operated on what are called interior lines of communication. They were surrounded on two sides (north and west) by the Union and the other sides by the Atlantic Ocean and Gulf of Mexico. But they were able to move troops and supplies within their area relatively quickly, while the Union shipments had to go around the long outer borders of the Confederacy. Railroads helped considerably. They also helped General Sherman cut through the heart of the Confederacy, a feat that would have been impossible without the railroads to carry troops and supplies.

Figure 6.4. *Constructing the railroad and its helper, the telegraph.* Here we see a telegraph construction crew working alongside newly laid railroad tracks in 1869. The two systems were mutually supporting. Compare with fig. 4.5. Detail of a photograph by Andrew J. Russell, courtesy Union Pacific Museum Collection.

The American transcontinental rail system was completed in 1869 when the Central Pacific and Union Pacific Railroads were linked at Promontory Point, Utah, in 1869. This was just eight years after the demise of the Pony Express, which would have been killed by the railroad had not the telegraph already did it in.

In the twentieth century electric power began to be used to drive some trains. They drew power from overhead wires. Another approach was to use a diesel engine in the locomotive that generated the electricity to drive the train. Eventually, the highest speed trains in America ran over 160 kilometers per hour and in France and Japan as fast as 320 km/hr.

But the replacements for the railways, automobiles, and airplanes came right behind them. By the mid-twentieth century, passenger travel by train was in decline and now lingers on in only a few places in the United States but may yet resurge due to overcrowded highways and airports and the new high speed trains that can beat an airplane in travel from city center to city center.

Wingless Aircraft[4]

In Greek mythology, the first person to fly was Icarus who made wings of wax and flew like a bird. Unfortunately, he flew too high, too close to the sun, and his wings melted. He became the first air crash fatality. Pegasus, also of Greek mythology, was a flying horse and may have represented a more pragmatic approach to flight by humans. Don't fly yourself, get some flying thing to carry you.

In real life, although Leonardo da Vinci (Italy, 1452-1519) designed some flying machines that never flew, the actually lifting of a human into the air was first done by hot air balloons. The first one that carried people aloft was built by the Mongolfier brothers in France, in 1783. A basic balloon is not steerable. Like a boat, the craft must be moving relative to the fluid (air) that supports it in order to be steered. The lifting power came from the fact that the warm air within the balloon is lighter than the cool air without, giving the name *lighter-than-air craft* to this class of vehicle. Early balloons were tethered to the ground. Such a vehicle served the French army for observation in 1794 during combat against the Austrians. They were also used for a similar purpose in the American Civil War and, untethered, for message carrying during the Franco-Prussian War of 1870.

A century after the Mongolfiers another French team flew a steerable airship powered by an electric motor-driven propeller. Airships (this term has stuck to designate lighter-than-air ships) were developed more or less in parallel with ground-based motor vehicles and were powered by electric or petroleum-fueled engines. At about the same time, Ferdinand von Zeppelin (Germany, 1838-1917) designed a rigid airship, a hydrogen-filled bag covering a metal framework (figure 6.5). Airships of his designs were used during World War I for observation and bombing. After the war, zeppelins were put into commercial passenger service. They could fly from Frankfurt to near New York in 52 hours, at about 110 km/hr, carrying 100 passengers. One such ship, the *Hindenburg*, caught fire and burned while attempting to land at Lakehurst, New Jersey in 1937, with great loss of life. This put an end to use of lighter-than-air craft for passenger travel and, as we shall see later, provided the first occasion for on-the-spot electronic news coverage of a disaster. The modern version of such airships, blimps, are filled with the inert gas helium. They were used by the U.S. Navy for antisubmarine patrol and are now mostly seen carrying advertising.

Internal Combustion[5]

In the mid-1800s Jean-Joseph Étiene Lenoir of Belgium built a two-cycle engine powered by a derivative of coal. In 1862 Alphonse Beau de Rochas of France built a four-stroke engine. Neither of these were practical. In 1878 Niklaus Otto, in Germany,

Figure 6.5. *The Graf Zeppelin*. Vessels such as this carried passengers and mail at about twice the speed of a fast ship. This one covered over a million miles in commercial service during its lifetime. Another zeppelin, the *Hindenburg*, suffered a disastrous fire in 1937, which put an end to this form of transportation The light object at the bottom near the front is the cabin containing passengers and crew. Photo © 2001 www. arttoday.com.

built a "successful" engine, and in 1886 Karl Benz (Germany, 1844-1929) built yet a better one and went on to found the automobile company today known as Daimler-Chrysler. Again, we see the pattern of different inventors in different countries gradually working toward the goal of a truly useful machine. These are called *internal combustion* machines. In a steam engine, a fire heats water, creating steam which is piped into a cylinder where it expands, driving a piston whose motion provides the power. In an internal combustion engine, the fire is inside the cylinder. A gas explodes, pushing on a piston, which again delivers the power where needed.

Automobiles[6]

Undoubtedly, the automobile's origin lay with the chariot, the first wheeled vehicle we know of. It was powered by animals but carried human riders. While they made their first appearance around 1000 BCE in Egypt, their drivers and owners must even then have yearned for larger, faster, and safer vehicles. Reaching this goal required two parallel developments, roads and motive power. Lacking these, Middle Easterners developed the camel saddle around 100 BCE. This enabled camels, "ships of the desert," to replace the chariot as a prime means of transport. Camels continued, as we know from art and movies, to be used for ceremonial and combat purposes.

Motive power remained in the hands, so to speak, of animals until the eighteenth century CE, when experiments began with steam engines. Steam worked well for ships and trains but not with cars. Yes, there was the famous Stanley Steamer automobile. It was in production for about the first twenty years of the twentieth century and set some speed records for the time but, in general, steam and cars did not fare well together. In the nineteenth century, good, workable internal combustion engines were being built, powered by fossil fuel, and more compact than steam engines. France, Germany, and the United States all lay some claim to the first automobile. Who really did it seems to depend on definition, so we'll leave the issue unresolved. But, around the beginning of the twentieth century, automobiles caught the public's fancy. There were many races, which tended to bring more publicity and more design competition. Speeds went from 18 km/hr, by a car built by Gottlieb Daimler (Germany, 1834-1900) in 1885, to 1021 km/hr (633 mph) by Richard Noble in 1983. Figure 6.6 shows an early Daimler car. Daimler joined with Benz to form the famous firm of Daimler-Benz that produced Mercedes-Benz cars. The current record was set using a "car" that was little more than a jet aircraft engine fitted out with wheels, not a vehicle for the highway.

Figure 6.6. *An early Daimler motorcar.* Gottlieb Daimler, the inventor, is chauffeured by his son in an 1886 car, among the earliest internal combustion powered automobiles. Photo courtesy DaimlerChrysler Classic Archives.

War, again, encouraged development. In World War I automobiles (actually taxis) were used to move 6000 French troops from Paris to the front for the Battle of the Marne, in 1914. Trucks carried supplies, one of them driven by the poet Gertrude Stein. The U.S. Army bought 17,000 trucks during the war. They served also as weapons carriers and evolved into the tank.

Shortly after the war, the United States raced to become an automobile-based culture. As early as 1922 more than 100,000 suburban homes were deemed to be wholly dependent on the car. By now, most American homes outside city centers, and many within, are surely in this predicament. The auto enabled rural free delivery of the mail, combined with electric trains and street cars to make suburbs feasible, and brought us the shopping mall.

Whether all these are beneficial is not clear. By now, we can hop into a car to visit anyone within a radius of about 3-500 kilometers (180-300 miles) in one day. We can send mail anywhere in the world and, although a letter may travel partway by plane or train, it will surely go at least partway in a truck.

In trying to assess the role of automobiles in communication, we must remember that they came about more or less in parallel with airplanes and after the establishment of railway trains. While it is certainly true that virtually any product delivered to a consumer travels at least part of the way in a motor vehicle, there is no clear measure of how much of a role these vehicles played in the communication of messages.

Airplanes[7]

The Wright brothers, Wilbur (United States 1867-1912) and Orville (United States, 1871-1948), worked at a time when several others were experimenting with *heavier-than-air craft*. The Wrights are generally credited with winning the race. A heavier-than-air craft, or airplane, is propelled through the air horizontally by a propeller (later by a jet engine) and is held aloft by air pressure pushing upward on its wings or other lifting surfaces. The Wrights' first flight was made at Kitty Hawk, North Carolina, on December 17, 1903. Figure 6.7 shows that flight. Slowly, the armed forces of various governments became interested, and by the time World War I broke out, some planes were available for use as observation platforms. Eventually, larger ones were used for bombing and more maneuverable ones for fighting with other planes. These fighter planes drew much romantic interest and the air battles produced some of the great heroes of the war—Baron von Richthofen of Germany, Billy Bishop of Canada, and Eddie Rickenbacker of the United States—who shot down the most enemy planes for their respective countries. A beneficial effect of all the aerial combat was a great improvement in performance of the craft.

Figure 6.7. *The Wright brothers' first airplane*. This was the first success-ful airplane, flown at Kitty Hawk, N.C., in 1903. Photo © 2001 www.arttoday.com.

During and shortly after the war, we had the first scheduled passenger air service (1914), which only flew across Tampa Bay, and the first airmail service (1918). The United States airmail services had carried 49 million letters by 1920. A group of U.S. Navy flying boats (airplanes that can take off from and alight on water) crossed the Atlantic Ocean in 1919 but required a stop-over. Charles Lindbergh made his famous solo, nonstop flight from New York to Paris in 1927. These flights brought tremendous public interest to aviation. In 1933 a Boeing airliner flew at a record 320 kilometers per hour. The Douglas DC-3 (figure 6.8) began service in 1936 as a passenger and freight carrier. It was the workhorse of commercial and military aviation through the 1940s and into the 1950s. It is still occasionally found in use.

Another Boeing plane, the 707, introduced in 1957, flew at a ground speed of around 1000 km/hr and the British-French Con-corde jet flew at mach 2, twice the speed of sound, or about 2400 km/hr. As this is written they have been grounded following an accident in 2000, but there is hope service can be restored. The fastest time recorded for an airplane has been 7313 km/hr by the U.S. X-15A-2 experimental aircraft.[8] By this time, ocean shipping as a mode of passenger travel or mail carrying was disappearing.

We can now fly from any major city airport to any other, anywhere in the world, within one day. Business travelers routinely set off from New York on Monday night and attend a meeting in Paris or London on Tuesday. But even this mode of communication is now being challenged by electronic communication media—sending messages instead of traveling to a meeting.

Figure 6.8. The Douglas *DC-3*. This plane was first built in the mid-1930s and became the mainstay of American commercial aviation through World War II and into the 1950s. It is still found in service today. Photo by Rick Radell, reproduced by courtesy of Mr. Radell and the Canadian Heritage Warplane Museum.

We have followed the story of transportation from the level of logs that fell into the water, to oxen pulling a cart, to jet planes crossing the Atlantic in little over three hours. From here on, the story of communication goes back to sending messages, rather than the messenger.

Notes

1. Coppin, *Cars, Trains, Planes*; Stowers, "Stationary Steam Engine"; Spratt, "Marine Steam Engine"; Field, "Internal Combustion Engine" are about engines. Landström, *Ship* is, obviously, about ships.
2. Landström, *Ship*.
3. Chandler, *The Railroads*; Clarke, "Building of a Railway"; Ellis, "Development of Railway Engineering"; Itzkoff, *Off the Track*.
4. Brooks, *Zeppelin*. Lighter-than-air craft, *Encyclopaedia Britannica*.
5. "Internal Combustion,"*Encyclopaedia Britannica*; "Steam Engines,"*Encyclopaedia Britannica* ; Field, "Mechanical Road Vehicles."

6. Angelucci, *Automobile*; McShane, Automobile.
7. Carpenter et al., *Powered Vehicles*; Gunston, *Chronicle of Aviation*; Gwynn-Jones, *Farther and Faster*; Sampson, *Empires of the Sky*.
8. When we talk about high-speed aircraft, flying at or above the speed of sound, we have to get a bit technical. Airplane speeds are given as *air speed*, or speed through the air. How fast the plane is going relative to the ground depends on how fast and in what direction the wind is blowing and on the density of air, which affects the speed of sound as do the air temperature and atmospheric pressure, which are in turn affected by altitude. Roughly, sound travels at 1,000 feet per second, or 680 mph. Mach 1 means one times the speed of sound; mach 2 is twice the speed, etc. Speed record: Funston, Bill. *Chronicle of Aviation*.

7

Telegraph[1]

Introduction

The idea for an electric telegraph almost certainly arose from the combination of semaphore systems, which had existed in one form or another for many years, coupled with increased scientific knowledge of electricity gained in the late eighteenth and early nineteenth centuries. As the nineteenth century dawned, electricity was still not sufficiently widely available, reliable, or reasonably priced for commercial uses. But things were happening.

In 1753 a letter published in *Scots Magazine* proposed the design of an electric telegraph. The idea of the writer, identified only as C.M. was to use twenty-six wires each connected to a different lamp at the receiving end. A pulse of electric current would be sent over the wire corresponding to a particular letter of the text of the message, turning on a lamp representing that letter.[2]

C.M.'s idea went nowhere at the time. It would have been cumbersome and expensive to rig so many wires over long distances. To light a particular lamp the sender would have had to send a single pulse down a selected wire. The code was the same for any letter: a pulse. What would have varied was the wire the pulse went on. C.M. seems not to have thought of the need for a symbol for the space, or end of word indicator, or numerals, or punctuation. In retrospect, his idea showed the immense value of what later became Morse code, allowing a single wire to carry a different code for each letter. Morse's code consisted of one to five symbols, either dots or dashes, which may have taken longer to send than C.M.'s pulse but required only one wire, making installation and maintenance so much easier. It was an example of the use of software to make hardware practicable. This difference is very important. We will see analogous developments when telephone is compared with telegraphy, or voice radio with wireless telegraphy. The first of each of these pairs is able to send more complex signals compared with the simple on or off of telegraphy.

After C.M.'s letter, but before Morse's work, Charles Babbage (England, 1791-1871) designed a mechanical calculator, called the analytical engine, that would use the position of a wheel or a tooth on a gear to represent a number. He was never able to build a working version himself but it was done later in Sweden. Joseph-Marie Jacquard (France, 1752-1834) used the predecessor of twentieth-century punched cards to record information to be fed to a loom in order to control the pattern to be woven into a fabric. His device controlled which threads of the warp were raised or lowered with each pass of the shuttle. In this manner, complex patterns that normally required a highly skilled weaver could be produced mechanically. We do not know whether or how Morse may have been directly influenced by these other inventors but, clearly, such ideas were in the air at the time.[3]

Beginnings[4]

Here are some brief sketches of early research scientists and engineers, working in electricity and magnetism, who had profound effects on telecommunication. Note how diverse they were in terms of nationality although limited to Europe and North America. Also included are some of the entrepreneurs who built industries out of inventions.

William Gilbert (England, 1540-1603) published the first book on electricity in 1600. In fact, he coined the word *electricity* from the Greek word *elektron*, meaning amber, a substance often used in early laboratories to generate static electricity.

Pietr van Musschenbroek (Netherlands, 1692-1761) invented the Leyden jar in 1745. This device for storing electricity was the precursor of the storage battery which was the power source enabling much of early electrical experimentation.

Alessandro Volta (Italy, 1745-1827) invented the Volta pile, or storage battery, and gave his name to a measure of electric potential, the *volt*. Batteries were needed at the time the telegraph was invented because there was not yet electric power in every home and office.

Hans Christian Oersted (Denmark, 1777-1851) observed in 1819 that an electric current affected the orientation of a compass needle, hence that there was a relationship between electricity and magnetism. This led eventually to the invention of the electromagnet, a key component of the telegraph and telephone.

André Marie Ampère (France, 1775-1836) formulated the physical law in 1825 that governs the relationship discovered by Oersted. His name survives as the name of a measure of electric current.

Michael Faraday (England, 1791-1867) discovered electromagnetic induction in 1831. This is the principle that a moving magnetic field will generate an electric current in a conductor within its field, the essence of an electric generator, or that a varying current could cause a magnet to move, the basis of the electromagnet.

Joseph Henry (United States, 1797-1878) showed in 1831 that an electromagnet could be made to ring a bell a mile away from the original source of the electric energy. He later became a partner of Samuel Morse.

Sir Charles Wheatstone (England, 1802-1875) was a professor at King's College, London. His name is associated today with the Wheatstone bridge, a device for measuring electrical resistance. He was an early developer of a successful electric telegraph.

Sir William F. Cooke (England, 1806-1879) worked with Wheatstone to construct a needle-based telegraph, one that used tapered indicators to point to an image of the letter being transmitted. This was patented in 1837 and is described later in this chapter.

Samuel Finley Breese Morse (United States, 1791-1872) was trained as an artist and was in fact a professor of art at New York University and a frequent traveler to England and Europe when he caught the "bug" of interest in telegraphy. He went on to build a successful telegraphic communication system and made it into a successful company.

Alexander Bain (Scotland, 1818-1903) was an early associate of Morse in the United States. While still in Scotland in 1843, he invented a device that could transmit pictures over a wire, the beginning of the modern facsimile (fax) machine.

Cyrus W. Field (United States, 1819-1892) was a businessman who became obsessed with building an undersea telegraph cable between Europe and America—and did it.

Emile Baudot (France, 1845-1903) developed a precursor of the modern system of encoding telegraphic and computer representations of letters, numerals, etc. His name survives in the name of a unit of signal transmission, the *baud*.

In addition to these key people, there were many others who experimented with the science of electricity or dabbled in inventions. Most never became famous or were first with anything. Work toward the electric telegraph was not that of a single man or team.

Telegraphs, Practical and Impractical[5]

S. T. Sommering, of Munich, developed a "chemical telegraph" in 1809. It used a separate wire for each letter, more or less as C.M. had proposed half a century earlier. The ends of the wires were immersed in water. When a current was applied to one end of a wire, bubbles formed at the opposite, immersed end, indicating which letter had been sent. This was intriguing but not highly practical. In 1823, in England, Sir Francis Ronald hung a cork ball from the end of each of a similar array of wires. The current created an electric charge on one ball, which was then repelled by the others, causing it to move. The moving ball indicated which letter had been sent. Then in 1825, Baron Schilling, a Russian diplomat on duty in Munich, developed a needle telegraph. An electric current caused a needle to swing left or right. A code was devised for letters, e.g., *left-right* was A, *right-right-right* was B.

William F. Cooke was a student at Heidelberg University at the time of Schilling's work. He brought Schilling's idea to Charles Wheatstone and the two worked to build an operational system that they patented in 1837. It, too, was a needle telegraph that used an electric current to deflect a series of needles at the receiving station. The needles pointed to a letter inscribed on the receiving device.

To the modern observer their system was quite complex. It involved five needles that could be deflected in such a way as to point to a letter or numeral printed on the face of a diamond-shaped equivalent of a clock face, shown in figure 7.1. Why not use various combinations of the five needles as a code, not as pointers? As we shall see, Emile Baudot devised a system based on such a code, but simplicity is not always first onto the scene. It can take inventive genius to achieve simplicity.

In the following year, 1838, the Cooke-Wheatstone telegraph was installed for the Great Western Railway between its stations at Paddington and West Drayton. The invention caused no great stir until 1845 when there was a murder in Slough, England, and a suspect was seen to board a Paddington bound train. His description was telegraphed ahead and as a result the culprit was apprehended, tried, and convicted. This could not have happened without a communication system capable of beating the train by at least the amount of time it took to summon police at the receiving end. It is not recorded whether the villain was aware of his place in history as he mounted the gallows.

There is evidence that a code system for letters was used in Greece as early as 200 BCE. The letters of the alphabet were arranged in a five-by-five array. Two sets of torches were used to signal the row and column of the chosen letter. This is much like modern communication and computer codes.

Figure 7.1. *The Cooke-Wheatstone telegraph.* This shows how the operator at the receiving end could see what letters were being sent. Electrical signals controlled movement of the various pointers. Imaginary lines through two nonvertical pointers converged at the letter being indicated, here V.

Samuel F. B. Morse had been visiting England at about the time of some of this early development work. He was, recall, an artist, not a scientist, but he expressed his dream in a letter written while at sea in 1832, returning to the United States:

> I wish that in an instant I could communicate the information, but three thousand miles are not passed over in an instant and we must wait four long weeks before we can hear from each other.[6]

After his voyage, he was supposed to have said to the ship's captain,

> Should you hear of the telegraph one of these days, as the wonder of the world, remember that the discovery was made on board the good ship *Sully*.[7]

Morse worked or consulted with Charles T. Jackson, Leonard D. Gale, Joseph Henry, Alexander Bain, Alfred Vail, and others. After much experimenting he produced a pen telegraph, or actually a pencil telegraph. Vail, both a technical associate and financial backer of Morse, was a cousin of Theodore Vail who would later be an associate of Alexander Graham Bell and head of the American Telephone & Telegraph Co. Like the needle invention in England, the Morse receiver used electric current to deflect an arm, this time holding a pencil, and thereby making marks on a moving roll of paper.

The coding of the marks made by this device indicated the letter being sent. The coding of letters is known to this day as Morse code, although there are several versions of it and some dispute as to who actually developed it. It is illustrated in figure 7.2. This method, using a single wire sending a different code for each letter, was a major advance in telegraphy. It may have required more skill by the operators but it simplified the mechanism, enabling it to be a more reliable machine. One form of a Morse telegraph transmitting key and an early receiver device are shown in figure 7.3.

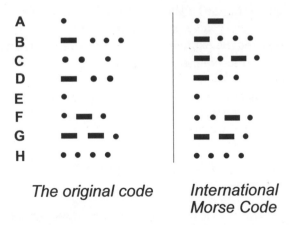

The original code   International
                         Morse Code

Figure 7.2. *Morse code.* A code using two basic symbols plus space or time between symbols to convey all the letters, numerals and some punctuation symbols for languages using the Latin alphabet. There are several versions. Shown here are an early Morse code and the widely adopted International Morse Code. The original version sometimes used spaces as part of a code, as with the code for C, and lacked ability to handle diacritical marks (as ñ, é or ü) needed in many European languages.

A U.S. patent for a telegraph was applied for in 1837 (note, the same year as Wheatstone and Cooke patented their system in England) but it was not granted until 1844. By that time, the developers had interested the U.S. government in becoming a user and sponsor. It took years to convince the government to award a grant, but once the money was available Morse's team built a telegraph line from Baltimore to Washington in only two months. On May 24, 1844, the famous message, "What hath God wrought," was sent from Baltimore with various members of Congress in attendance at a demonstration in Washington. (Incidentally, that sentence is usually written with a question mark, but it is a biblical quotation (*Numbers*, 23:23), written as an exclamation, not a question, and surely seems an exclamation in this context. There was no telegraphic symbol for punctuation at that time.)

Figure 7.3. *Morse telegraph components.* On top is a telegraph key or transmitter, simply an electric switch. When the round button at left is pressed, a circuit is closed, allowing current to flow. When released, a spring causes the circuit to break. Below is a receiver, essentially an electromagnet. When current is received the magnet pulls one piece of metal against another, making a clicking sound. Both date from the late nineteenth century. Variants of the receiver are a pencil pulled by the magnet onto a moving paper tape or a buzzer. Reproduced courtesy of the Collingwood Museum, Acc. Nos. X976.241.3 and X976.241.2

After this demonstration things developed quickly. The telegraph, still to be improved in many ways, was clearly practical. One change made by Morse was to make the receiver sound-based, using a buzzer, a method that was faster and more reliable than the pen-based methods. The operator heard the code and wrote down the corresponding letter. There was also interest in a true printing telegraph, one that printed the actual letters rather than codes, and there was interest in finding a means to use the wires for more than one message at a time. Royal E. House (United States, 1814-1875) built a true printing telegraph—one that printed letters of the alphabet—which he patented in 1844, just a few months after Morse's successful demonstration in Baltimore. What seemed to be a superior means of telegraphy, House's system required two men at each end to operate it because it needed power beyond what a simple battery could provide. The men had to operate a treadle mechanism to power the system.

In 1858 Wheatstone invented a machine that could read codes recorded in a punched paper tape and then transmit them at ten times the speed of a human operator. The tapes had to be prepared in advance, of course, but this method was quite useful in sending out news dispatches. Some variation of this was used until computers came into use for busy telegraph offices.

Next came multiplexing, a method of sending two or more messages simultaneously over a single line. When this was first done is another of those questions with multiple answers. Wilhelm Gintl in Vienna in 1853 devised a method of sending two messages at once (duplex). Others also tried, but J. B. Stearns in the United States is credited with the first successful duplex transmission in 1871. Thomas Edison extended the capability to four messages in 1874.

In 1874 or 1875 Emile Baudot devised a method of sending five on-off signals at once.[8] These were sent sequentially but triggered by a single action of the operator, not multiplexed. A keyboard having five keys was used, several of which could be depressed simultaneously. They were held momentarily as a distributor scanned the keys, determined which were in use, and transmitted the corresponding five-element code showing which keys were depressed and which were not.

Five two-state symbols can be combined in 32 ways to produce codes for letters of the alphabet and functions such as the carriage return. Baudot did not use a code of all *off* symbols, leaving only

31 possible codes. Two symbols were reserved to indicate *start numbers* and *start letters* which had the effect of creating a second set of meanings for the other keys, one called letters, one called numbers. This is more or less what we have with modern typewriter and computer keyboards, using the shift key. The number keys were used for punctuation as well as numerals. This code enabled operators to use a simple keyboard to send and receive messages. They did not need the special training of a Morse operator. The Baudot code, or a variant of it, shown in figure 7.4, was used until quite recently in teletype transmitters. By extending the number of what came to be called *bits*, we arrive at the coding now just about universally used in computers.

Letters	Figures	1	2	Keys 3	4	5
A	1	●				
B	2			●	●	
C	3	●		●	●	
D	4	●	●	●	●	
E	5		●			
● ● ●						
Figures					●	
Letters						●

Figure 7.4. *The Baudot code.* An operator depressed from one to five keys on a keyboard, then all five basic symbols were combined into a five-bit character showing which keys were depressed and which were not and then automatically transmitted. The five-element code is similar to that used in modern computers. *Figures* means that codes to follow represent figures until the *Letters* code is sent.

As the telegraph became practical, investors began to set up commercial companies to provide telegraphic service. Not too surprisingly, money was a chronic problem and this led the developers to enter new partnerships repeatedly to continue their commercial development.

Much of this development was happening in the mid- to late nineteenth century in the United States when two other important events occurred. Railroads were being built at a great rate, taking over from canals, and extending into new inland territories where canals could not easily be built. Railroad operators needed to know where their trains were, especially as many lines had only single tracks. Then, the Civil War came along with its great need for rapid communication that only the telegraph could provide.

During that war, Gen. William T. Sherman led his devastating march, referred to earlier, through the heart of the Confederacy, capturing Atlanta, then continuing to Savannah and the sea, then through South and North Carolina. All this played a major part in ending the war. He paid effusive compliments to railroads and the telegraph.

> . . . [T]hat single stem of railroad, four hundred and seventy-three miles long, supplied an army of one hundred thousand men and thirty-five thousand animals for the period of one hundred and ninety-six days. . . the Atlanta campaign was an impossibility without these railroads. . . . For the rapid transmission of orders in an army covering a large space of ground, the magnetic telegraph is by far the best. . . . Hardly a day intervened when General Grant did not know the exact state of facts with me, more than fifteen hundred miles away as the wires ran."[9]

The combination of railroad industry needs and a government at war created great markets for the new communications medium and helped to establish what became a great industry.

As we said, Morse did not actually invent the telegraph. That came from England. His erstwhile partner Vail accused him of not having contributed anything to the *science* of telegraphy, and Vail, depending on the source you read, either actually invented Morse code or significantly improved on the first version. But Morse made it all work, made it practical.

How It Works

The essence of Morse's telegraph was that it transmitted a pulse of electricity from one place to another over a wire and varied the length of the pulse. Morse's dash was three times the duration in time of a dot. This was a big improvement over the Cooke-Wheatstone model that required multiple lines to encode a letter but still not as elegant as Baudot's yet to be invented code and encoder. What the telegraph could not do was to send a tonal message. Codes were more efficient than yes-no signals but were not yet a tuned sound that could reproduce the human voice. For that, we had to await the telephone, not much longer in coming.

The telegraph worked by converting an electrical pulse into sound, light, or marks on a paper. But the sound, if used, would be monotonic. Information was conveyed by varying the number or duration of the sounds.

Telegraphy Becomes a Business[10]

We have already seen that Samuel Morse was not alone in doing development work in telegraphy. We remember him, but not most of the others, because he not only built a working machine, he found adequate financing and built a business around it. He went into partnership briefly with Alfred Vail. Together, Morse and Vail managed to secure a grant from the U.S. government. This was not the first instance of government sponsorship of technological development, which became the common mode by mid-twentieth century. But the first company to be set up in the business of transmitting and delivering telegrams was funded by Amos Kendall with Morse as a partner.

James Gordon Bennett, Jr. (born Scotland, 1795, emigrated to United States, died 1872), the publisher of the New York *Herald,* was among the first users of the telegraph for news reporting only four years after the Baltimore demonstration. He led the way in forming the Associated Press to pool the use of the telegraph by the *Herald* and other newspapers. In 1851 the New York and Mississippi Valley Telegraph Company was formed from a number of smaller firms which, by 1856, became the Western Union Telegraph Company (WUTC). The newly constituted firm was led by Hiram Sibley (United States, 1807-1888). It came to dominate the industry in the United States until it virtually disappeared under pressure from the telephone and later, computer-based electronic mail.

In 1861 a transcontinental telegraph line was completed from San Francisco to Omaha and then to Chicago, from which there were already lines to the East Coast. This was only twelve years after the discovery of gold in California and seventeen years after the device was patented in the United States. This was also the year of the start of the Civil War after which the railroads expanded explosively. Gradually, telegraph wires were strung along railway tracks to the mutual benefit of the two industries (figure 6.4). The telegraph companies could place their poles along the railroad right of way, affording easy access to them for installation and maintenance and also providing service to a prominent customer. The railroad companies could be kept informed of where their trains were, critical when trains ran both ways on a single track.

The new Western Union expanded quickly and wanted to provide service between America and Europe. Before a transatlantic cable was successfully laid, Sibley began exploring the possibility of running a line through Alaska, to Siberia, then on to western Europe. While the Atlantic cable eventually made this proposal unnecessary, American interest in Alaska was kindled and contributed to its purchase from Russia in 1867.[11]

While the telegraph never became a household appliance, it was commonly used in business, especially in stock market trading where the ubiquitous Thomas Edison devised a form of telegraph used to bring up-to-the-minute stock prices into traders' offices. Telegraph service tended to be used by families only for special occasions: sending congratulations, condolences, or emergency messages. For the common person the telegraph was not a hands-on or interactive device. Fast as it could transmit, messages had to be taken to the telegraph office and from there to the recipient's home or office. The stock ticker, a printing telegraph, could of course be read directly but it was a one-way system. No on-line stock purchases in those days.

The industry hit its peak traffic-carrying load between the World Wars. It went into sharp decline after World War II, not directly because of the war, but because the war had increased the need for communication facilities when materials were scarce and the telephone companies, also limited by shortages, could not keep up with the demand for message traffic. When they did catch up after the war, putting two-way, easy-to-use communications in the great majority of North American homes and offices, it was to seriously injure the telegraph industry. In the 1960s computer to computer communication, usually over telephone lines, was becoming a major factor in the telecommunications industry.

A derivative of Western Union still exists primarily for transferring money. Money is "wired," not of course, by actually transmitting money, but by sending a message from one telegraph office to another authorizing cash to be delivered to a named recipient. A similar arrangement was developed for arranging to send flowers over long distances, originally called Florists' Telegraph Delivery. The message specified what was to be delivered, to whom, and from whom, but the receiving-end florist took the flowers from local inventory. These were early versions of what we now call e-commerce or electronic commerce.

The telegraph industry represents one of the few examples of a major communications medium being almost totally displaced by another medium. Of course, telegraph itself had replaced the Pony Express. Usually the loser changes in some way, but does not disappear, as we shall see in the case of television's effect on radio.

Crossing the Oceans[12]

Let us go back to 1849. The telegraph had been used in England and Europe since 1837 and in the United States since 1844. Steam locomotives and steamboats had been invented, but it was not yet a high-tech world. Electricity existed and was in use but was not available for industrial power and certainly not for home use. A voyage across the Atlantic could take several weeks, so trans-oceanic business transactions took a considerable amount of time and provided no guarantee of safe passage.

Yet it was in that year, 1849, still relatively early in the saga of electricity, that the first underwater telegraph cables were laid, one crossing the English Channel, one in the United States crossing the Connecticut River. There were all sorts of questions about how to insulate a wire that sat for a long time in salt water, how to protect the cable from the dragging anchors of ships, over how long a distance could an electric pulse be detected, how even to carry the huge weight of cable needed to cross an ocean and pay it out at a steady rate from the ship carrying it. There were no Yellow Pages in which to search for companies that could supply the appropriate insulation material or armoring to protect the cable, although rubber, in a crude form then called *gutta percha*, was just beginning to be used commercially and was of great help in insulating cables. Because of this lack of materials and experience, early attempts tended to fail, as did the cross-channel effort. In 1855 the British eventually were successful in laying a cable across part of the Black Sea, during the Crimean War.

The American Cyrus Field got the fever around 1853. About that time an attempt was made to run a telegraph cable from Newfoundland south to the United States. The purpose was to allow incoming transatlantic ships to stop in St. John's and send forward important messages to the U.S., saving several days transit time. Even this was not successful, so anyone who thought he could lay a cable thirty-two hundred kilometers across the ocean from England must have seemed insane. But enough people saw the potential value that money was made available.

The U.S. and British governments collaborated by lending naval ships to the effort and, after many tribulations, a cable was completed in 1858. It failed very shortly thereafter. Another one was laid in 1865. It failed. Finally, in 1866, success!

It may be something of a measure of the magnitude of this accomplishment when we realize that the first transatlantic *telephone* cable was not completed until 1956. The first live transatlantic television broadcast occurred in 1962.

The Industry Goes Downhill

We mentioned the decline in traffic after World War II. This was largely due to improvements in telephone technology and to more and more people becoming used to having the communications instrument in their own homes or offices. Also, they could now afford a telephone. They could even afford long-distance calls, which have long been the prime source of telephone industry profits.

This idea of the end user being the operator, or "hands-on" user of equipment has proved irresistible. The same phenomenon occurred later in computing when the personal computer became more popular than anyone initially predicted. One factor in the personal computer's popularity has been electronic mail, or e-mail. This, in effect, makes every computer user a telegraph operator. It continues a trend begun by House and his printing telegraph, Edison with a telegraph machine in the office of stock traders, and then the telephone that gave hands-on use but not printed records and, until very recently, could not store messages for later pickup by the recipient. So, we might consider that, rather than disappearing, the telegraph industry has grown in popularity, but only by use of a different technology and in the hands of different companies.

What may be the final chapter in telegraphy as Morse knew it came on January 1, 2000, when the United Nations' International Maritime Organization decreed that oceangoing ships must begin

use of a new system of signaling called the Global Maritime
Distress and Safety System, which automatically sends out distress
messages as well as the location of the sending ship. Ships using
this are no longer required to monitor the radio frequency over
which distress signals have traditionally been sent using Morse
code. The code is not banned, but will probably disappear from use
in the maritime world.[13]

The electric telegraph was the first of what many think of as the
telecommunications media. Certainly it was the first based on
electricity. There were various semaphore or light signals before,
but they depended on someone repeating the signal again and
again, from each station to the next, in order to reach far distant
places. The repetitions cost time, money, and reliability. The
telegraph spanned the entire length of the wire or cable in what
seemed to be no time at all—crossing the ocean in a fraction of a
second.

While it did not enter the home or most small businesses and
was not a tool of the average person, it established the idea that
high-speed, long-distance communication was here to stay. Once
communication at this speed became routine, the world was never
the same again.

Notes

1. Coe, *Telegraph*; Garrat, "Telegraphy"; Reid, *The Telegraph in
America*; Shiers, *The Electric Telegraph*; Standage, *The Victorian
Internet*. An excellent and extensive bibliography can be found on
Neal McEwen's Web site, "The Telegraph Office," (address in the
bibliography).
2. C. M., "Expeditious."
3. "Jacquard," *Encyclopaedia Britannica.*
4. Coe, *Telegraph*; Garrat, "Telegraphy"; King, "Development";
Shaffner, "History"; Taton, *General History.*

5. Coe, *Telegraph*; Garrat, "Telegraphy"; Thompson, *Wiring a Continent*.

6. Coe, *Telegraph*, 27.

7. Coe, *Telegraph*, 27.

8. Guillemin, "Telephone Apparatus."

9. Sherman, *Memoirs*, 888-891.

10. Israel, *From Machine Shop*; Lubrano, The Telegraph; 40; Thompson, *Wiring a Continent*; "The History of Western Union."

11. "The History of Western Union"; Sherwood, *Exploration of Alaska*, 16.

12. Clarke, *How the World*, 3-105; Field, *The Story of the Atlantic Telegraph*.

13. Nickerson, "For Ships" is a newspaper account of the change in communication procedures. Kent, "The Global Marine" is a technical description of the new, automatic replacement for Morse code distress signals.

8

Telephone[1]

Introduction

The telephone was one of the most significant inventions in history. It ranks with writing and printing in terms of the extent of its rapid and widespread adoption among the world's population and its influence on society. It had three capabilities that the telegraph did not have: it carried the human voice, it could easily be used by almost anyone without special training, and it was interactive; when you spoke someone answered right away, not by a return telegram perhaps received the next day.

As with other major inventions, the telephone did not spring full-blown from a single person's work, although in common with other inventions, a single person got most of the public credit. That person is, of course, Alexander Graham Bell. He lived in a litigious age and had to defend the priority of his invention often in the law courts.

In the Beginning Was the Word[2]

The question for the early telephone workers was how to transmit
the full range of sound of spoken words, instead of monotonic
sounds, over something like a telegraph. There were the usual
pioneers who had worthy ideas but never quite came up with a
working telephone. Probably because of the appearance of similar-
ity of telephone to telegraph apparatus, there was a good bit of
migration of workers from one to the other.

Sir Charles Wheatstone, the telegraph pioneer, invented what
he called an *enchanted lyre* in 1821. He linked the sounding boards
of two musical instruments by a pine wood rod. A tune played on
one was reproduced on the other. The rod carried the message but
this did not involve electricity and could not travel very far nor
reproduce the human voice. It transmitted sound acoustically, that
is, the sound waves went through the wood and were never converted into electricity.

In 1837 an American, Dr. C. G. Page, created sound by moving
an electromagnet in a magnetic field. He called the result "galvanic
music" but went no further toward a telephone.

Charles Boursel of France is credited with first suggesting, in
1854, that sound could be transmitted electrically by using a
diaphragm whose sound-induced vibrations made and broke
contact with an electrode as the waves impinged on the diaphragm.

Johann Reis of Germany developed an early transmitter in 1861
based on Boursel's idea. He could convert sounds to electric
current, transmit them, and reconstruct the sound at the other end.
But these were rather pure tones, not complex mixtures of tones as
are made by the human voice.

Physicist Hermann von Helmholtz, (Germany, 1821-1894) did
some more fundamental work and demonstrated that vowel sounds
could be transmitted, but again not the full complexity of the sound
of a human voice.

The Telephone Is Born[3]

Alexander Graham Bell (1847-1922) was born in Scotland, emigrated to Canada and then to the United States. His father, Melville, was an instructor in elocution. The elder Bell devised a method of recording the positions or movements of the throat, tongue, and lips in order to make various sounds. He called it *visible speech*. The notation was intended for students of foreign languages and people who were deaf and wanted to learn to speak. Graham, as the son was called, or was it Aleck? Sources disagree. had two brothers both of whom died from tuberculosis and this, plus the death of Bell's mother, caused the father to decide to move the family to the more healthful climate of Canada. Graham's mother had begun to lose her hearing when he was around twelve years old and this created in him an interest in teaching the deaf, which was in fact his main occupation when he began to work on electro-mechanical ways of creating or re-creating speech.

To this day, Canadians claim that the telephone was invented in Canada because much of the development work was done there. But none doubt that the famous first telephonic voice message, "Mr. Watson, come here. I want to see you" was uttered after an accident in a laboratory in Boston, Massachusetts. (The exact wording varies with the person reporting but the essence is always the same.) This was on March 10, 1876. The date is important because his filing for a patent was done on February 14 and a patent was issued on March 7 of that year, both *before* the thing really worked.

As with the telegraph, the ability to take a new invention from lab to market was crucial and Bell, like Morse before him, was able to do this, with help.

Sic Transit Gloria Mundi[4]

The title above, for non-Latin readers, translates as, "Thus goes the glory of the world." Bell's glory still lasts, but Elisha Gray (United

States, 1835-1901) lived and worked at the same time as Bell, perhaps contributed as much or even more to telephony, but is largely unheralded today. He invented several telegraphic devices and was cofounder, with Enos N. Barton, of a company that evolved into the Western Electric Company. Unlike Bell, he had a strong background in electrical theory. He worked on the idea of a telephone at the same time as Bell and got the idea for a transmitter that was better than the first one Bell proposed to the Patent Office. He filed what is called a caveat with the Office, a warning that he was at work on an invention which he planned to try to patent when complete. This filing was also on February 14, 1876, but a few hours after Bell filed. Some think that Bell got to see this caveat and thereafter amended his patent filing to include a similar transmitter. Ever since, there has been a question as to who really invented the telephone. What we know for certain is that Bell filed first and was awarded the patent.

Gray did not end up impoverished. In fact, Western Electric was at one time the largest single component of the American Telephone and Telegraph Company, scion of Bell's original company. But, neither did he acquire the fame or riches of Bell. Today there are forty-two books in the Library of Congress about Bell and none about Gray. His colleagues believed he had invented the telephone. "Dr. Gray's associates maintain, and . . . nearly all the scientists in the country concede, that to his brain is attributable the invention of that idea which has fairly worked a great revolution in telegraphy (*sic*)."[5] Gray's Western Electric Company lived on until very recently when it became part of Lucent Technologies, Inc. His name, though, is usually raised today only to point out the importance of inventors filing for patents as soon as possible. He missed the grand prize by a whisker and the prize for second place is unhappily very different from that for first.

The Patent Issue[6]

Let us take a closer look at the situation regarding that patent, beginning with what a patent is.

A patent is a license granted for a limited time by a government to an inventor in order to prevent others from making and selling essentially the same product that the inventor had produced first. Patents and copyrights were provided for in the United States Constitution, so important did its authors think they were. Article I states that, "The Congress shall have Power . . . to promote the Progress of Science and useful Arts, by securing for limited Times to Authors and Inventors the exclusive Right to their respective Writings and Discoveries. . . ." The idea is that a person invests time and money to create some new thing. This thing, be it a song or a machine, will enrich the society into which it comes and its developer deserves the right not to have others simply take it from him. In general, more devices being patented makes for a more vigorous economy and encourages further innovation.

The inventor files with the Patent Office a claim that something new has been produced. The Office must verify that nothing like it has been filed before—priority of filing was the basis for turning down Gray—and, if not, a patent may be awarded. An award can be overturned if a new claimant can prove prior discovery; difficult but not impossible to do. As we shall see, one of Marconi's early radio patents was overturned years after it was granted.

We can't get too technically detailed here. Bell's patent application was entitled "Improvements in Telegraphy." It did not claim the ability to carry speech but it did claim the ability to carry some sounds. Remember that the telephone does not literally carry sound, but a *representation* of sound, but we talk colloquially as if it does.

Bell's original transmitter was quite crude; Gray's was better. A marginal note in Bell's application proposes a Gray-like transmitter as an alternative approach. No one knows when that marginal note was added, but it was believed by some not to have been in the body of the original document. Since Gray was the developer of the method that was proposed, that formed the basis for Gray's challenge to the Bell patent. In brief, Bell won, not only over Gray but over others including Thomas Edison who challenged him.

Sound and Electricity[7]

We should introduce a little physics here. This section can be skipped by readers who know the subject.

Sound, recall, travels in the form of waves. The waves are created by some object vibrating at an appropriate frequency. The human ear can detect sounds in the range of about 20-20,000 vibrations or cycles per second. Frequency is measured in these terms. One cycle per second is called one Hertz (1 Hz). The actual range of what can be heard varies among individuals. The waves are successive compressions and rarefactions of air (or some other substance, such as Wheatstone's pine wood) caused by the vibration. You see a similar effect when you drop a pebble in water and watch the waves emanate from the point of impact (figure 8.1). Familiar sounds come from striking a drumhead, plucking a guitar string, striking a tuning fork, or vibrating our vocal chords.

Wavelength is the distance from the crest of one wave to that of the next one (figure 8.2). All sound waves travel at the *same speed* in any given medium, say air or water, but at *different speeds* in different media. In any medium the waves *attenuate* or lose power as they pass through it, but the denser the medium, the farther the waves can travel. You may have experienced putting your ear to a railroad track, thereby hearing a train's sound through the steel rails long before you could hear it through the air.

Since sounds go at the same speed in any medium, what varies in order to produce different pitches is the wavelength. The shorter the wavelength, the more waves can pass a given point in a second. That number, the number of waves per second, is what is called the *frequency*. Sopranos sing at a higher frequency than basses but both generate sounds in a wide range of frequencies. This difference between the simple tonal structure of the tuning fork and the complexity of the human voice is what made the telephone difficult to invent. It is relatively easy to transmit sound at a given frequency. It is much harder to transmit a number of frequencies simultaneously. Figure 8.3 shows a pure tone, all sounds of the same wavelength and frequency, such as a tuning fork or single plucked guitar string might generate.

Figure 8.1. *Making waves in the water.* Dropping an object into still water causes a series of concentric waves.

Figure 8.4 shows the multiple frequencies generated simultaneously by a human voice. Here, distance from left to right indicates the passage of time. Distance from bottom to top indicates increasing frequency of the sound waves generated. Darkness at any spot indicates the intensity or power of the sound generated at the corresponding moment and frequency. A violin or piano also produces a range of simultaneous frequencies because the box or sounding board adds some tones, making what we hear a richer sound. An instrument like the triangle is virtually a tuning fork, producing more of a single-frequency sound.

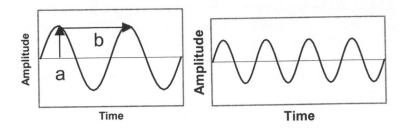

Figure 8.2. *Basic wave struc-ture*. The amplitude of a wave is the height above the zero level, indicated by **a**. (If water, the zero level is the surface of the water.) The wavelength is the distance from a point on one wave to the corresponding point on the next, **b**.

Figure 8.3. *A pure wave form.* This is the wave pattern gener-ated by something like a tuning fork that produces sound waves of uniform amplitude and fre-quency, although the amplitude will eventually attenuate and fall to zero.

   The graph of figure 8.4 is called a *voice spectrograph*. It is difficult to read in any detail but it does give an idea of the com-plexity of the problem, how to get a transmitter and then a receiver to vibrate with all the necessary frequencies at once in order to reproduce the speaker's voice. That was the challenge that faced Bell and his competitors.

   An important concept in telecommunications is *bandwidth*, essentially the amount of information that can be sent at one time over a transmission medium. It is analogous to the diameter of a pipe carrying fluid. More water can flow through a sewer pipe big enough for a person to crawl through than through a half-inch copper tube. As we'll see, more information can flow through a fiber optic cable than through a copper wire. Much of the develop-ment work in communications in the twentieth century has been a quest for more bandwidth.

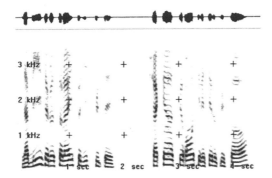

The voice, speaking in French, said, "La bise et le soleil se disputaient," then, "Chacun assurant qu'il était le plus fort." In English, "The sun and the wind were arguing, each assured it was the strongest."

Figure 8.4. *A voice spectrograph.* An instrument as complex as the human voice generates sound in a number of frequencies simultaneously. Here, the horizontal axis represents the passage of time as words shown to the right were spoken. The vertical axis represents the frequency of sound waves generated and the relative darkness indicates the intensity or power generated at any frequency. The inset at the top shows the maximum and minimum wave heights at any time. Photo courtesy Dr. Philippe Truillet and Prof. Paul Milgram.

How the Telephone Works[8]

The simplest form of a telephone system requires a *transmitter* to convert sound into electricity, a transmission *channel* consisting of a copper wire to carry the electrical signals, and a *receiver* to receive the signals and convert them back into sound. Such a system would allow an originator of a message at A to talk to a destination at B and vice versa, provided both parties had both a transmitter and a receiver. This is the way the system worked at the very beginning.

Older readers may remember making a telephone out of two tin cans or round boxes (oatmeal boxes were best because they were cylindrical) connected by a length of string attached to the bottoms of the cans as in figure 8.5. If the string was kept taut, words spoken into one box could be heard in the other. The sound waves vibrated the bottom of the speaker's box, vibrating the string, which vibrated the bottom of the other box and recreated the sound. Like Wheatstone's enchanted lyre the transmission did not involve electricity. Each box could serve alternatively as transmitter or receiver. Sound quality was poor, range was limited, and the string often broke.

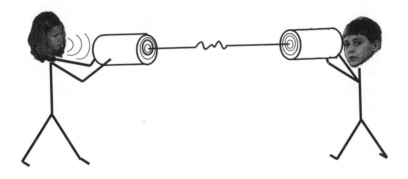

Figure 8.5. *A primitive sound-powered telephone.* Many a child has made a telephone system like this one. Each can has a string or wire attached at the bottom, usually by boring a tiny hole in the bottom surface, pushing the end of the string through it, and then tying a knot to keep it in place. When the line is stretched taut, words spoken into one can cause the bottom to vibrate, in turn vibrating the string, in turn vibrating the bottom of the receiving can. The receiver's vibrations regenerate the original sound—almost. There is a considerable loss of sound fidelity.

The earliest telephones worked like the cereal box phones but used electricity to carry the sound rather than a vibrating string. Like the string telephone, there had to be a line, now electric wire, going directly between any pair of people wishing to communicate with each other. It was not long before it became apparent that to connect each *telephone subscriber*, as users became known, with every other one would require a great deal of wire. It would have been simply impractical; wires everywhere. Hence, the idea formed of having each subscriber communicate with a central location that could then connect the incoming call to the appropriate destination's line. This center is known as a *switching center* or *exchange*. Let's consider these components, one at a time.

Figure 8.6. *The first patented Bell telephone.* This is the instrument Bell first patented. It served as both transmitter and receiver. Photo courtesy AT&T Archives.

Transmitters and Receivers

The function of a telephone transmitter is to convert sound waves to electric current in such a way that it can be transmitted over a wire to a remote receiver. (For those of you who know that radio is often used as the carrier, wait. That comes later.) The receiver's function is the opposite, it converts electric current into sound. One of Bell's early systems (figure 8.6) used the same device for both transmitter and receiver, but we will describe a transmitter developed by Thomas Edison in 1877 that is fairly simple in concept.

Start with a cup made from a solid piece of carbon (figure 8.7). Fill it with granulated carbon and then place a sort of plug on top and cover the opening so as to contain all the carbon. Run an electric wire from the carbon cup to the plug. Then, place a thin sheet-metal diaphragm over the opening.

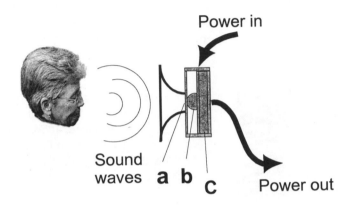

Figure 8.7. *Schematic diagram of a telephone transmitter.* This is the type of transmitter Thomas Edison developed. Sound waves press on the metal diaphragm **a,** which presses on a carbon button **b**, which in turn compresses carbon granules **c**, and that compression affects the electrical conductivity of the carbon, hence changes the electric current flowing through the device, indicated by the power in and out lines.

As you hold the device near your lips and talk, the sound waves make the diaphragm vibrate, just as with the cereal box. The vibrations cause the carbon granules to be alternately compressed and restored to original shape. This compression changes the resistance of the carbon to an electric current passing through it and the changing resistance causes variation in the flow of electricity. So, by starting with a constant flow of current, then varying it in accordance with the sound the speaker makes, we achieve an electric current that is an analog of the sound. And that's the job of the transmitter.

The receiver is a little less of a Rube Goldberg device. It uses an electromagnet (figure 8.8) to move a diaphragm. An electromag-

net is an instrument that is magnetized only when electric current
flows in wires wrapped around an iron core, and the magnetic
field's strength varies with the amount of electricity. So, the
variable current that came from the transmitter changes the magnetic field generated by the electromagnet and that, in turn, attracts
and releases the diaphragm, re-creating sound waves.

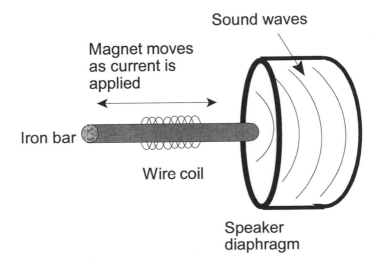

Figure 8.8. *Schematic diagram of a telephone receiver.* Electric current
from the sending telephone causes the magnetic field around the wire coil
to vary, exerting a variable force on the iron bar which then presses
against the diaphragm, generating sound waves. It is something like a
telegraph receiver (fig. 7.3) only instead of a pencil or sounder being
either up or down, the diaphragm moves to a variable extent, thereby
delivering a more complex signal.

If you question whether so simple a system can reproduce
sound like your stereo can, you're right to do so. The telephone is
not, or was not, a high-fidelity instrument. In fact, it does not, even
today, reproduce the entire frequency range of human voices. But,
telephones were invented to carry human *speech*, not the opera.

(Actually, one of the early uses *was* to carry scheduled news and music, in broadcast form, i.e., going out simultaneously to all subscribers. This was done in Budapest in 1893. Sound quality would have been minimal but early users were apparently enchanted that they could get any music at all.)[9] New ways of encoding the sound can improve this, such as using digital representation, but for the first hundred years or so, telephone users got along with slightly distorted voices.

Lines

Connecting two telephones is not as simple as stringing a length of copper wire between them. Today, much of telephone communication is carried out using glass wires or radio in the form of microwave. For now, let's stay with the original methods, which were in use until the middle of the twentieth century.

We can't do with just a strand of wire. There needs to be a circuit, that is a loop to which both sending and receiving phones are connected. Fortunately, we can use the ground as part of the loop. Even if you're not familiar with electricity, you will surely have heard about grounding an appliance. So, string a wire from one phone to the other, then a wire from each phone literally into the ground, and you have a circuit.

The next problem is how to tell the other party you want to talk. You can't expect all telephone users to stay constantly glued to their phones waiting for a call. There has to be a way to signal that a call is coming. Further, the power needed to send messages came from batteries in the early days, and users did not want to use up battery power keeping the circuit closed all the time. (A closed circuit is complete, it has no breaks. An open circuit has a gap somewhere, preventing current from flowing.) In some early phones, the receiver was suspended from a hook (figure 8.9) attached to the main works and this hook was attached to a spring, in turn attached to the main circuit which, in its normal condition was open. When you lifted the receiver, the hook popped up, closing or completing the circuit. You would then turn a crank on your telephone, causing a signal to be sent to the other phone, ringing a bell. The other person picked up his or her receiver, closing the circuit at that end, and it was now possible to talk both ways.

Figure 8.9. *A nineteenth-century telephone "on hook."* When the receiver or ear piece is lifted, a spring pushes the hook up and, inside, causes a circuit to be closed that connects the telephone instrument to the exchange. Today, we usually have a button or other protuberance under the handset serving the same function. Photo by C. T. Meadow. Telephone from the collection of Linda and Ian Samuels.

Telephone transmissions can be affected by lightning and other sources of electromagnetic emissions—noise—that impinge on the lines. To help counter this, the phone companies use two wires twisted together. By some clever processing of the signals the combination, called a *twisted pair*, can reduce the effect of the noise.

Stringing telephone lines looks easy and is not conceptually difficult but it can be physically difficult as well as expensive. Lines have to cross rivers and scale mountains. They have to be protected against lightning, squirrels gnawing at them, the weight of winter ice pulling them down, or of wind or errant trucks

Figure 8.10. *Telephone lines abound in snowbound New York City.* This picture was taken during a severe blizzard in 1888. More astounding than the snow is the number of lines, only twelve years after the telephone was invented. Photo courtesy AT&T Archives and the Museum of the City of New York.

knocking down the supporting poles. The poles also have to be protected against rot. Figure 8.10 shows a scene in New York City, during the devastating blizzard of 1888. Remembering that this was only twelve years after the first telephone patent, the number of lines evident seems remarkable. It is of little wonder that microwave radio, buried lines, or use of glass fibers that could carry many more messages through a lighter transmission channel was so attractive to telephone companies.

Whether a line is going across town or across the Atlantic Ocean, the signal is going to attenuate and has to be amplified or repeated at intervals. This can add significantly to the cost of any line or cable. Typically, a repeater is needed for about every mile of conventional cable.

Considering the intensity of interest and investment in a transatlantic telegraph cable, it is interesting that a transatlantic telephone cable was not completed until 1956. The big problem, by comparison with the telegraph cable, was the much greater need for amplifiers than were needed for telegraphic signals, in order to preserve intelligible speech. Also, the availability of radio lessened the need for a phone cable, but when it was finally available, the cable offered greater bandwidth than radio, allowing more calls to be carried at one time.

Switches and Exchanges[10]

To have to string a wire between every pair of telephone users was not an attractive prospect. It would have been economically impossible long before cities became completely covered over with telephone wires. It took only until 1878 before a new arrangement was worked out. Each member of a group of subscribers was connected with a central station, the exchange. The caller told an operator at the exchange which other subscriber he or she wished to call and the operator connected the incoming call to an outgoing line connected to the destination phone.

This meant that each subscriber needed only one line to the exchange and it is still that way today. We are still describing plain old telephone service, or POTS. Wireless telephony is described later. At the exchange there has to be a way to connect the incoming line with the line to the destination's phone. This was done simply by linking the two lines with a connecting wire. A typical switchboard (this came much later than 1878 but performed the same functions) is shown in figure 8.11. The operator hears the caller identify the destination, sends a signal that a call is coming, and when the destination answers, connects a wire from the originator's line to the destination's line, completing the connection. If the destination subscriber is busy or not answering, the caller would be so informed.

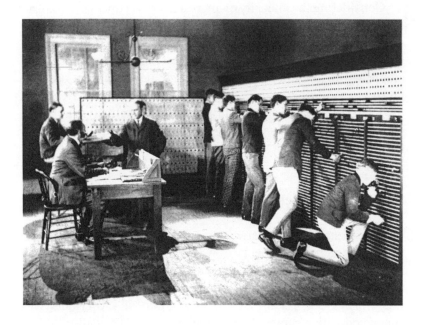

Figure 8.11. *An early switchboard in 1879.* Originally, the "operators" were boys. Because of their tendency toward boisterous behavior, they were soon replaced by women who were used almost exclusively until mid-twentieth century. Again, note the size of the board and the number of people required to operate it only three years after the telephone was invented. Photo courtesy AT&T Archives.

Originally, the caller identified the destination to the operator by naming the desired destination but in 1879 telephone numbers were introduced. The need for them is another indication of how quickly telephoning caught on. Operators at first were boys but their often unruly behavior soon led to their replacement almost exclusively by women. Whichever, they were still talking to the caller in the 1930s, although dial telephones began to be used in the 1920s. You picked up your receiver (went *off hook*) and then heard a voice say, "Number please." By the end of the 1940s dial telephones were in use over most of the United States.

In 1891 Almon B. Stowger developed an automatic switch-board. He was an undertaker and was said to have been suspicious that operators were routing calls intended for him to his competitors. The automatic switch worked only among the subscribers connected to a single exchange and was not widely used at first. Today, of course, we can place a call to almost anywhere in the world by direct dialing. This means crossing oceans, mountains, and international borders even between countries not otherwise on cooperative terms with each other. We can connect to a telephone system that may be technically different from our own. We can even place calls to anywhere from an airplane. The total accomplishment is really astounding.

In order to make connections with far distant phones, more is necessary than to contact the local exchange and provide a number. That exchange must be able to contact other exchanges, possibly many in sequence, before the final connection is made to the remote destination. Typically, in the North American system, a local exchange can hand off calls directly to other local exchanges or can contact an exchange in the same city that handles long-distance calls. That long distance (or in AT&T parlance *long lines*) center can connect with other long-line centers in other cities, but the basic nature of the connection is the same as at the local exchange. And so, we have evolved a giant, worldwide network of telephone systems (figure 8.12).

Setting up a long-distance call means making all the linkages from origin to destination and holding them all for the duration of the call, including for periods when no one is actually talking.

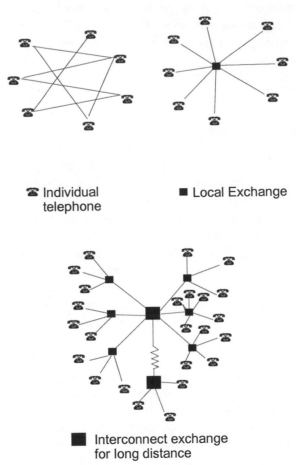

Figure 8.12. *Telephone networks*. At top left is a depiction of a network in which every telephone is wired to every other one that the owner may want to talk to, which was the original method. Such a configuration would be impossible today, with millions of telephones in use. The second illustration (top right) shows a group of phones connected to a single exchange, each requiring only one wire that goes only to the exchange. The third network (bottom) is more like today's. Individual phones connect to exchanges, which connect to other exchanges, creating a gigantic, world-wide network. The connections are no longer only by wire.

When either phone goes on hook, that is a signal to the system to break all the connections and free the equipment for other calls.

Telephone as a Business[11]

While Bell was doing what we now call development work, he was still teaching deaf students. The fathers of two of these students, Thomas Sanders and Gardner Greene Hubbard, became interested enough to provide Bell with some financial support. The three signed an agreement in 1875 to create the Bell Patent Association, each member to share equally in the benefits from any patents. On July 9, 1877, when Bell was all of twenty-three years old, the Bell Telephone Company was formed to supercede the Patent Association. Elisha Gray and Thomas Edison were right on Bell's figurative heels, but the Bell company was without peer in its ability to make a practical product and successful business out of the telephone. In the fall of the year of its founding, there were over 600 telephone subscribers, even without exchanges yet having been set up.

While this company was growing, Gray and Western Union were challenging the Bell patents. They didn't win but they did divert energy and money. As one result, in 1878 more capital was needed and several new partners were brought in. They, in turn, brought in Theodore N. Vail (United States, 1845-1920) to run the company. He had begun his working career as a telegraph clerk and was to go on to play a dominant role in the telephone industry well into the twentieth century. In 1880 the American Bell Telephone Company was created to concentrate on operations and licensing of patents.

Also in 1878 the number of Bell telephones in service reached 10,755. In 1879 Western Union, which by then owned Western Electric, gave up its patent battles with Bell. In 1881 American Bell bought Western Electric from WUTC.

In 1884 long-distance service was available between Boston and New York. This was to become a major factor in the industry. It is an expensive undertaking to install a telephone in a home or office. It requires stringing a line from home or office to the nearest exchange, running wires into the interior of the building, placing a telephone instrument therein, and maintaining all this equipment. But it only required one cable, with several lines, going from an exchange in Boston to one in New York and passing through several intermediate cities that could also use the lines. This meant a relatively smaller investment per call carried.

Until the 1970s AT&T owned all the equipment in its service areas. They would not sell a telephone, only rent it to customers. It was Vail's idea to provide only a complete, packaged service. The equipment was a part of the package, not to be sold separately. They provided a reliable service with a single point for customers to bring any complaints. The objection was mainly price. In the 1970s it was antitrust law suits that forced AT&T to abandon this policy.

Once most homes and offices had telephones, the big source of new revenue came from long-distance service. Charges for local telephone service were fixed, per subscriber, and did not depend on the number or duration of local calls. But charges for long distance were based on the duration of the call, so the more calls that could be encouraged, the more revenue that resulted. This only required linking exchanges, not a triviality, but less expensive than linking all individual subscribers.

Foreseeing the growth and benefits of countrywide service, American Bell created a new subsidiary, called American Telephone and Telegraph Company (AT&T), a New York Corporation. New York was closer to financial markets.

The success of the Bell companies did not go unnoticed. Shortly, competition grew up and by 1897 there were over 5000 independent telephone companies in the United States. In that year they formed a trade association, the National Independent Telephone Association. Although Bell and its successor companies were generally thought of as monopolies, these independents continued to exist and at least one, General Telephone, grew quite large.

By 1900 there were a million telephones in use in the United States. Figure 8.13 shows the growth of the number of instruments or access lines in the United States over the period 1876-1997.[12] As the twentieth century dawned the telephone business was big, still expanding, and hungry for capital. To help raise money, American Bell, a Massachusetts company, transferred its assets to New York-based AT&T to make sale of shares easier. AT&T then became the head of the empire. Theodore Vail, who had been president from 1878 to 1887 became president again in 1907 and stayed on until 1919.

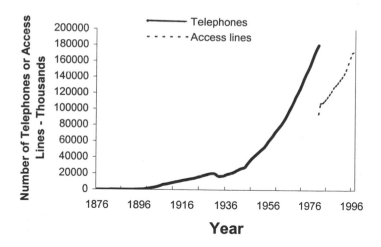

Figure 8.13. *The number of telephones or lines in the United States, 1876-1996.* We see here a dramatic rise in the number of telephones in use, particularly after World War II. Around 1980 these statistics began to be reported in terms of access lines, rather than number of instruments. This reflects an increase in the number of multiple-phone homes and in the use of computers attached to telephone lines. There may be far more instruments than lines in any home or office.

Mr. Bell had always maintained a laboratory for new developments. By 1925 the lab was incorporated as a subsidiary company
half owned by the parent AT&T and half by another subsidiary,
Western Electric. The new entity was called Bell Telephone
Laboratories, one of the most productive commercial research
laboratories ever. It produced a raft of electronics inventions and
three teams of Nobel Prize winners.

For nearly a hundred years it seemed like triumph after triumph
for AT&T but explosive growth in anything cannot go on forever.
Eventually troubles appeared, the result of a series of law suits,
some brought by government, some by private companies. One of
the first was brought in 1948 by the Hush-A-Phone company. They
made a device that attached mechanically, but not electrically, to
a telephone to protect the privacy of the speaker by covering the
mouthpiece. Since Bell owned all equipment used by their customers they did not want any "foreign" devices to be attached. They
tried to block use of the Hush-A-Phone. But the upstarts sued for
the right to attach and eventually won. Shortly thereafter, another
relatively small company, Carter Electronics Corporation, also
sued to get rights for their customers to attach a non-Bell device to
a Bell phone. They made the *Carterfone*, a device used to connect
a radio with a telephone, which *was* an electrical connection. They
won also, and thereafter it became common for other companies to
make equipment that attached to Bell telephones or directly to the
phone line.

One of the biggest blows came from Microwave Communications, Inc. (MCI). In the 1960s a number of Bell's long-distance
customer companies complained that Bell was not providing the
service they needed for frequent, private, long-distance communication among a company's geographically dispersed offices and
plants. MCI applied to the Federal Communications Commission
for permission to operate a microwave service between St. Louis
and Chicago and they eventually won that right. They did not stop
there; they are now, under the name MCI World Com, one of the
largest competitors in the long-distance telephone market.

In 1974 the U.S. Department of Justice brought an antitrust suit
against AT&T. It took until 1982 for the two parties to reach a

settlement, to take effect in 1984. This provided for the divestiture by AT&T of its operating subsidiaries. AT&T could no longer be in the local telephone service business. They were allowed to retain their Long Lines Division, Western Electric, and Bell Laboratories, but the twenty-two Bell subsidiary operating companies were to be split off and grouped together as seven independent companies, called "Baby Bells,"[13] which were no longer required to buy their equipment from Western Electric. On the other hand, the new firms were not allowed to manufacture; they were service companies. Gone, finally, was Vail's idea of a complete package of service and equipment all provided by a single company. Gone was a virtual AT&T monopoly in long-distance service.

What was the effect of the breakup? Most people feel it was for the best. Customers saw competition develop and with it lower prices and new ideas. Stockholders of the venerable bluest of the blue chip companies found themselves owning shares in eight corporations, instead of one, and they all did well. The split-off companies have since combined with other companies or split themselves. They have gone into businesses other than local telephone service, such as cable and satellite television transmission.

New Developments

In its earliest days, the telephone was a telegraph-like device that transmitted messages from an origin to a destination that was connected to the origin by wire. The difference between the two systems was that the telephone could transmit most of the range of sound frequencies that came from a human voice, while the telegraph, if used for sound, sent only one frequency. Gradually, technology was added to permit any phone to be connected to an exchange and from there to any other phone linked to a network of exchanges. Telegraph could do the same but was never extended into the home or to virtually all persons in a business or office.

Once telephone exchanges were created, attention was focused on automating them, to allow the caller to dial a destination's number and not rely on a human operator to make the connections. AT&T, its successors, and competitors never stopped adding new capabilities and services to the basic telephone system. We can only survey these briefly.

Basic Components

While the quality of sound and reliability of the system continued to improve, nothing more of a dramatic nature came after the invention of the automatic switches until after World War II. During that war, *microwave radio* began to be used for communication. This will be described in more detail in chapter 9, but for now it means that beginning in 1950, high-frequency radio, not just heavy metal cables, could be used to carry telephone messages. This was particularly important in crossing rivers and oceans, mountains, and even parts of built-up cities.

After microwave came *fiber optic cable,* beginning in 1979. This is also described later.

Another use of radio has been in *cellular telephones.* A cellular phone is really a radio. There are sending-receiving stations, similar to telephone exchanges, throughout most of urbanized North America and other parts of the world. These stations have a range of around one mile. An outgoing call goes to the station, then, possibly through other cells, to an exchange that connects with the telephone network. As the caller moves, perhaps in a car, from one cell to another, the call is passed off to an adjacent cell. Figure 8.14 shows some antennas for cellular phones mounted atop an apartment building, a sight we are seeing more and more often. The main advantage of cell phones lies not in their small size but in the lack of need for wires, cables, or fibers. By using satellites the range of cellular telephones is greatly extended. The systems that use satellites also user higher frequency radio waves and have tended to be called *personal communication systems* (PCS) but it's the cellular idea, enhanced quite a bit. PCS is or can be worldwide in coverage. See more detail in chapter 11. Figure 8.15 shows a modern wireless telephone.

Figure 8.14. *Cellular telephone appears on the scene, literally.* These antennas sit atop an ordinary apartment building, serving the cellular telephone industry. Urban rooftops can now be revenue-earning real estate for the owners. Photo by C. Meadow.

Figure 8.15. *A wireless telephone.* Today's telephones are small enough for pocket or purse, unencumbered by external wires, and able to reach into the worldwide telephone system from almost anywhere through a local cell. This one weighs only about four ounces. The cells are found in most populated areas in most industrial countries and are popular in developing countries as well. Photo by C. T. Meadow.

Figure 8.16 shows a schematic of how cellular calls are routed, from the caller to a cell antenna, to a central station, then to the cell in which the recipient is currently located. To know where the recipient is requires that the phone system frequently contact each of its subscribers to determine in which cell they are found. This contact is not apparent to the owner of the telephone.

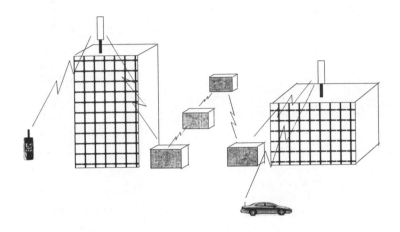

Figure 8.16. *Routing of a cell-phone call.* The message goes from the caller's cell phone to a nearby antenna, then to a computer in the cell, which routes the call to a computer containing information about the destination's location. Then it goes to the antenna nearest the recipient and finally to the destination's phone, here, in a moving car.

Automation[14]

The earliest automation of the telephone system goes back to Stowger's automated switch in an exchange. It replaced the human operator in accepting an incoming call, ascertaining what connection was desired, and connecting the originator's line to the destination's. But it took abut fifty years before direct dialing was in full use, and another twenty or so before direct-distance (long-distance) dialing was available. Beginning in the 1960s other automated features were introduced:

Voice mail in a sense began as attached answering machines but now may be handled by telephone company central equipment. Its basic functions allow a subscriber to leave a message for callers in case of no answer or a busy signal and allows the caller to leave a message to be heard later.

Call waiting allows a person engaged in one call to see or hear an indication that another person is trying to reach that line. The person called can switch back and forth between the two calls. Irritating to some (e.g., your author) it is nonetheless ever more popular.

Call display provides a visual display at a receiving telephone of the telephone number from which an incoming call originates and possibly the caller's name. This gives the potential recipient advanced knowledge of who has to be dealt with and gives the option not to answer.

Redial, after a busy signal, offers the caller the option to have the telephone system monitor the called number, then when it is free, dial it again and call back the originator.

One number means that if you own several telephones, say a fixed one at home and a cell phone in the car or pocket, callers may call a single number and the system will try the alternate lines in sequence if there is no answer.

Call block allows a recipient to block calls from certain originating numbers, such as telemarketers whose numbers were detected using call display. (It is interesting that the same companies that offer this service also do so much to support the telemarketers trying to reach us.)

Privacy or Its Lack

Most of us grew up thinking of the telephone as a completely private instrument. No one could hear what we were saying other than the person at the other end of the line. Wiretaps change this to some extent, but they require, at least in the western democracies, some form of judicial authorization. Our neighbors or business competitors cannot listen in on our calls. But portable and cellular phones, remember, are radios. The content of the calls made through them goes out as a broadcast signal that can be intercepted. British Prince Charles suffered considerable embarrassment when a romantic call was picked up and made into the day's news. Credit card numbers have been intercepted from cellular phone calls. For most of us who know the range of these phones is limited and that a listener would have to put in a great deal of effort, we don't worry about the problem. But it is there.

The heavy use of telemarketing bothers many people. While not exactly a security intrusion, it is often seen as a violation of privacy, especially since calls so often come during the evening mealtime, or children's bedtime. Even call display means that the destination of your call knows who is calling before you get to announce it, not a major issue for most people, but disquieting for some. It represents a change in the way we have to regard our telephone communications.

Fiber Optics

In the early days of telegraph and telephone, it was a great struggle to erect poles and string lines over nasty terrain or to lay the first undersea cables. For the people who did this the idea that a thin glass fiber could replace large numbers of copper wires would have seemed astounding. Further, the glass fibers are immune to interference from other electrical lines, magnetic fields, or thunder storms.

Glass fibers carry information in the form of light waves, not electricity. A typical glass fiber telephone line can achieve transmission rates of 10^{14} (a hundred million million) bits per second, while the traditional copper wires leading into the home or office are limited to 28.8 thousand bits per second.[15] These numbers depend on how data are encoded and how wires are protected from noise. The glass is of higher capacity, lighter in weight, and less bothered by external interference. It can carry a signal 150 km without the need for a repeater. Why not use it everywhere? Only because it is not yet installed everywhere and installation to every subscriber's home or office would be expensive. If all local lines, from subscriber's location to telephone exchange, were converted to glass fiber the local-service telephone companies would leap over the television cable operators in terms of bandwidth they could make available to the subscriber.

Fiber optic cables are proliferating in long-distance applications, especially undersea cables. These have come to compete with satellites as the inexpensive, reliable means of carrying messages across a continent or to other continents. We'll have more to say about this in chapter 11.

Competition

With the breakup of AT&T and the development of wireless telephone and cable and satellite television, there was a wild scramble to restructure the telecommunications industry. Telephone companies merged with cable television companies and cable companies competed for Internet connection service, originally mainly routed through the telephone companies. New technology changes the value of old at a dizzying rate. One example is that the Internet can be used for voice transmissions, allowing voice calls that do not use the telephone system directly. We have not yet covered the Internet, but most people have at least heard of it. The advantage to the user of going through the Internet for voice calls is price—it is or can be much cheaper to make long-distance calls this way.

What we are seeing and will discuss more fully in chapter 13 is that communication services are tending to encroach on, merge with, or converge upon each other. From this might develop, if not complete chaos, a highly versatile communication system that can handle voice and other sounds, data messages such as accounting data, and pictures, whether still or full-motion. Distance will gradually disappear as a major factor in cost of transmission and eventually in price to the consumer.

Notes

1. Brooks, *Telephone*; Fischer, *America Calling*; Grosvenor, *Alexander Graham Bell*; Pool, *Social Impact*; Stern & Gwathmey, *Once upon a Telephone*.
2. "Telephone" in *Encyclopaedia Britannica*, 9th ed.; Brooks, *Telephone*.
3. Stern & Gwathmey, *Once upon a Telephone*.
4. Brooks, *Telephone*, 43-49
5. Stern & Gwathmey, *Once upon a Telephone*, 10.
6. Brooks, *Telephone*, 47-81.
7. "Basic Concepts of Wave Theory," *Encyclopaedia Britannica*.
8. Brooks, *Telephone*; "Telephone and Telephone System," *Encyclopaedia Britannica*.
9. Marvin, "Early Uses of the Telephone."
10. Brooks, *Telephone*.
11. Brooks, *Telephone*; Crandall, *After the Breakup*; Danielian, *AT&T*; Irwin, *Telecommunications America*; Temin and Gal-ambos, *The Fall*.
12. Data for this figure came from several sources: Bank, *Cross-national*; *Historical Statistics of the United States*; Mitchell, *International Historical Statistics; Statistical Abstract of the United States*.
13. The seven (one almost wants to say dwarfs, but they turned out to be anything but) were: Bell Atlantic, BellSouth, Southwestern Bell,

Ameritech, US West, NYNEX, and Pacific Telesis. They have since gone on an acquisition and name-changing binge.
14. A good general review of these new features is found in the Toronto white pages telephone directory and probably that of other cities as well.
15. Held, *Understanding Data Communication*, 691.

Part 4

Electronics

Our story is now about to enter the modern age, that of electronics and what may in the future be known as the age of information. It began at the beginning of the twentieth century. Electronics has since brought us a level of information transfer and processing capability unimaginable in 1900 and, even more, to a recognition of the importance of information in all aspects of life. Information has *always* been important, even in the days when it was only used by primitive people to warn of an attack or tell the location of a good fishing ground. In those primitive times there would not have been much information and certainly none that needed to be instantly transmitted thousands of kilometers away.

What is the difference between electricity and electronics? It is not sharply delineated. Electricity, as a field of study, is concerned with the flow of electrons, usually through solid conductors such as copper wire. It gave us the telephone, telegraph, and electric lights. Older dictionaries define electronics as concerned with the flow of electrons through a vacuum or a gas. It is also concerned with electromagnetic waves. When vacuum tubes were introduced, they were used to convert alternating current to direct current and to amplify faint currents by varying the flow of electrons through the vacuum. Remember the importance of amplification to under-sea telegraph cables and their need for frequent repetition or

179

amplification of currents grown faint from attenuation. Electronics has given us radio, radar, television, and computers.

Modern electronics also deals with *solid state* mechanisms, beginning with transistors. These pass electric current through various solid materials, such as silicon, rather than gasses, achieving much faster and smaller circuits that produce less heat. These gave us modern computers, televisions, cellular telephones, pocket-size radios, microwave ovens, laptop computers, and such other wonders as electronic games and micro-listening devices. But the important thing is that electronics is mainly used for transmitting or processing information, not for lifting or driving heavy objects as were steam, internal combustion, and conventional electric current when used to power machinery in the nineteenth century.

Electronics has brought many real benefits to humanity. One is that ships at sea and airplanes in flight can contact the ground to summon aid in case of emergency. Another is that news of all kinds, not just of marine and aviation disasters, flows quickly and we are able to be better informed. Still others are the raft of new diagnostic and treatment devices developed for medicine, a whole new way of providing news and entertainment, and conducting commercial transactions. Electronics has virtually taken over entertainment through radio, television, cinema, photography, and now the Internet. Not everyone sees automatic bank machines as a gain for civilization, but some do because they allow communication with the bank at all hours and from many locations.

On the other side of the coin, there is now nearly permanent storage of personal data that we may not want stored and the ability for prying eyes (or fingers) to find bits and pieces of information about us filed in diverse places. Nor is everyone delighted with the saturated coverage of riveting news, such as shootings at a school, that we get from television.

Is the final judgment on all this technology going to be favorable or unfavorable? One answer is that, marvelous as these things are, they are still machines; they must be properly controlled by their inventors and users.

We shall start this narrative with radio, the first great new product of the electronic age. Radio has many uses, notably wireless telegraphy, broadcasting, radar, and carrier service for telephones, television, and computers. We'll also describe the invention of the transistor and integrated circuits, which spurred the development not only of modern radio but of all other forms of electronics. Then, we go on to television, communication satellites, and computer to computer communication, all of which make use of radio technology.

9

Radio[1]

Introduction

The telegraph and telephone freed us from having to travel or send someone to deliver a message over a long distance, except under limited circumstances in which some form of semaphore could be used. Radio started us on the process of cutting the umbilical cord to the wire network. The phrase "wired city" became a catch phrase in the late twentieth century. It means that more and more machines in homes and offices were connected to each other. It is ironic that, as this expression began to gain currency, we were beginning to move away from our dependence on wire and toward increased dependence on wireless radio, in its various forms, to link ourselves with television broadcasters and telephone centers.

Beginnings[2]

We are going to see a pattern, similar to that of telegraph and telephone, of a set of people from different countries working separately, each making a contribution, until one of them, building on what others have done, invents something that combines all this progress into a working, practical, and profitable product. Part of the pattern is a series of legal squabbles over patent rights. We will see that, after the earliest developments in Europe, contributors tended to come more and more from North America and that, regardless of origin, most were formally trained scientists. It became ever harder to acquire the requisite knowledge without the help of a university. Just as with the telephone, many people contributed to radio but one name stands out as having gotten and deserved most of the acclaim and riches—in this case Guglielmo Marconi. Here are the other major players of the earliest days of radio.

Michael Faraday, mentioned in chapter 8, showed that an electric current produced a magnetic field around itself, the current induced the magnetic field, hence the term *induction*. This principle is used in the electromagnet, important in telephony, illustrated in figure 8.8.

James Clerk Maxwell (Scotland, 1831-1879) demonstrated that it was possible to detect changes in an electrically-induced magnetic field at a distance. He also showed that the induction was reversible, that a moving magnetic field could induce an electric current, and he believed that the effect traveled in straight lines at the speed of light. This effect came to be known as *electromagnetic waves*. He formulated Faraday's discoveries into precise mathematical terms.

Heinrich Hertz (Germany, 1857-1894) confirmed Maxwell's predictions and built what were probably the first radio transmitter and receiver. His name was later given to the measure of wave frequency.

Sir Oliver Lodge (England, 1851-1940) may have beaten Marconi to some important discoveries but did not concentrate on ways of applying them. He later acknowledged that he did not foresee their importance to the military and merchant marine. He won U.S. Supreme Court recognition of his accomplishments after his death.

Guglielmo Marconi (Italy, 1874-1937) experimented with the production and detection of electromagnetic waves at ever increasing distances. His father was Italian, his mother Irish, of the family that produced Jameson's Irish whiskey. His early scientific education and experimentation were done in Italy, but his mother's family had contacts in England where some of his most important experiments were conducted, his patents first awarded, and his business formed.

More Physics

It will help our understanding to add a bit to our miniphysics lesson of chapter 8. We're going to cover more about bandwidth, the electromagnetic spectrum, frequency allocation, and the range of radio signals.

Bandwidth

If you had some kind of transmitter capable of sending only one wave per second you could not get much information across in that second. How much can you vary a single wave to make it a meaningful signal? The only choice would be to vary its amplitude or signal strength. Make that two Hz and you can at least send one loud and one soft, in either order, or both loud or both soft. Jump up to 1000 Hz (1 kilohertz or 1 kHz) and you can begin to sense how many different symbols—combinations of wave variations—you can send in a second. If you are going to transmit music over a low-bandwidth channel you will probably lose some of the higher and lower tones. Increasing the bandwidth enables you to keep more of them, resulting in higher fidelity music reproduction. This accounts for the popularity of FM broadcasting for music. Bandwidth is actually measured as the difference between the highest and lowest frequencies in use. If you have a transmitter operating in the range 90,000-100,000 Hz, the 10 kilohertz (10 kHz) difference is the bandwidth. Electronics engineers are constantly striving for more bandwidth. It's the modern version of gold. Anytime you can devise a way to increase the bandwidth of an existing communication system, such as the telephone network, that really will put lucre in your pocket.

The Electromagnetic Spectrum

Electromagnetic waves appear in many different forms. Visible light is one form, so are X-rays, radio waves, and ultraviolet light. The difference among the forms lies in the wave frequency or the wavelength. Since all electromagnetic waves travel at the same speed in a given medium, the higher the frequency, the shorter the waves must be. Given a measure of either, you can easily compute the other. The speed is about 300,000,000 meters (186,000 miles) per second. At a frequency of one Hz, it would take one second for one wave to pass a point and the wavelength would be an enormous 300,000,000 meters. In the frequency band used for AM radio, say one million Hz (1000 kHz or one megahertz, 1 mHz), the wavelength is 300 meters. In the region used for television and FM radio, the frequency might be ten million Hz and the wavelength three meters. X-rays have frequencies such as 10^{18} Hz (1 followed by 18 zeros), with wavelengths on the order of three ten-billionths of a meter.

Figure 9.1 shows some of the major types of electromagnetic waves and their frequencies and wavelengths. Radio began using the lower portion of what are now called the radio frequencies but has been edging upward, into higher frequencies ever since, giving higher bandwidth.

Frequency Allocation

If all the radio transmitters, say in one city, used the same frequency, we would have a great mess. No one would hear anything clearly. Music would be mixed with taxi dispatching, the sports news with the fire department. The principal method of keeping order so that this does not happen is frequency allocation, assignment by a government agency or international organization of different frequencies for different users. There are some exceptions. You may have heard one taxi interfering with another, because all the cabs for any one company and their dispatcher may use the same frequency. Discipline is required to avoid chaos.

Wave length (m)	Frequency (Hz)	Type
3×10^{-15}	10^{23}	Cosmic rays
3×10^{-14}	10^{22}	
3×10^{-13}	10^{21}	
3×10^{-12}	10^{20}	
3×10^{-11}	10^{19}	X-rays
3×10^{-10}	10^{18}	
3×10^{-9}	10^{17}	Ultraviolet light
3×10^{-8}	10^{16}	
3×10^{-7}	10^{15}	Visible light
3×10^{-6}	10^{14}	
3×10^{-5}	10^{13}	Infrared light
3×10^{-4}	10^{12}	
3×10^{-3}	10^{11}	
3×10^{-2}	10^{10}	Microwaves
3×10^{-1}	10^{9}	
3×10^{0}	10^{8}	TV and FM radio
3×10^{1}	10^{7}	AM radio
3×10^{2}	10^{6}	
3×10^{3}	10^{5}	
3×10^{4}	10^{4}	Long wave radio
3×10^{5}	10^{3}	
3×10^{6}	10^{2}	
3×10^{7}	10^{1}	
3×10^{8}	1	

Figure 9.1. *The electromagnetic spectrum.* Electromagnetic waves display quite different characteristics depending on their wavelength or frequency, ranging from very long wave radio to very short wave cosmic rays. Visible light and X-rays are other examples of this phenomenon.

Transmission Range of Radio Signals

We know that visible light reaches us from galaxies so far away
that the distances are hard to imagine and that it comes through
empty space. Radio waves originating on earth travel through air
which has the inhibiting effect called *attenuation*. Further, high-
frequency radio waves travel in a straight line, hence they can only
reach a point on earth as far as the eye can see—to the horizon.
The distance depends on the power and on how high above ground
the transmitting and receiving antennas are, but a typical FM
station has a range of about 50-80 km or 30-50 miles. Thus, it is no
surprise that radio transmitting antennas are put in high towers or
on top of high buildings to reach out a little farther. Toronto's CN
Tower, at 553 meters (1815 feet) tall, the world's tallest unsup-
ported structure can get signals out to around 85 km at ground level
or over 100 km to a tower. Differences in range gained by height
of towers is illustrated in figure 9.2.

Lower frequency radio can go further. Very low frequencies
follow the curvature of the earth. Some higher frequency waves
can reach long distances by being bounced off the ionosphere, a
layer of charged particles in the upper atmosphere. These waves
can be reflected off the ionosphere, back to earth, up again, and
down again, as in figure 9.3. This is how transoceanic radio works.
These techniques are much at the mercy of general atmospheric
conditions. You may have experienced hearing an AM station at
night, from quite far away, only to have it fade in and out as you
try to listen.

Wireless Telegraphy[3]

Back, now, to the main story. Marconi, beginning in the late 1890s,
successively sent and detected electromagnetic waves across a
room, then between buildings, then over 1700 meters, then 56 km
across the English Channel, and eventually across the Atlantic

Ocean. What he had was a wireless telegraph. It could send and receive electromagnetic waves, convertible into sound without tonal variation, but which could be encoded just as Morse and others had done with ordinary electric current. It was nowhere near ready for voice transmission but there was to be plenty of demand for what he had.

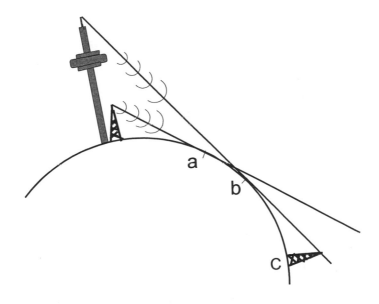

Figure 9.2. *Line of sight transmission.* High-frequency radio waves go in a straight line to the horizon. This is why broadcasters use high towers and why you cannot normally pick up an over-the-air television broadcast from more than about 80 km or 50 miles away. While this picture is not to scale it shows how a higher tower, such as Toronto's CN Tower, can add to the distance transmitted, adding about 45 km to the average range. It also shows the advantage of a high receiving antenna.

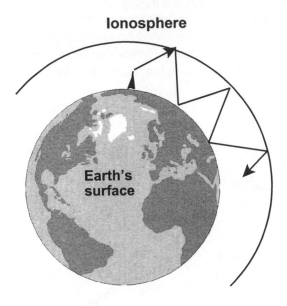

Ionosphere

Earth's surface

Figure 9.3. *Extending the range of waves by bouncing off the ionosphere.*
It is possible to extend the range of high-frequency waves by bouncing
them off a layer of charged particles above the earth, perhaps several
times.

In order to reach even the 1700 meter distance, Marconi had to
create a means of amplifying the signals, a theme that will be
repeated in any discussion of radio. He first accomplished this in
1894 with what he called the *coherer*, to be attached to an antenna
made of electrically conductive material such as copper. The
antenna picks up electromagnetic waves "in the air," creating a
very slight current because, as Faraday demonstrated, a moving
magnetic field generates a current in an electrical conductor. See
figure 9.4. The coherer consisted of a number of tiny bits of nickel
arrayed randomly in a glass container. Upon detecting an electric
current these bits aligned themselves, becoming coherent and able
to conduct and amplify electricity. If the container were then struck

by a small hammer, the pieces would lose their coherence, becoming again a jumble of nontransmitting fragments. When the antenna finishes passing on a signal, it would trigger the tapping hammer to reset itself. See figure 9.5. At this point Marconi was convinced he could make wireless really perform in practical applications.

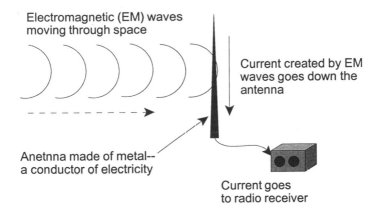

Figure 9.4. *Magnetic induction.* Faraday discovered that a moving magnetic field induces an electric current in a conductor within the field. Radio waves are variations within a magnetic field. They induce a current in a receiving antenna which leads to the receiver as in figure 8.8.

As usual, there was the need to demonstrate the new machine, to interest the public and potential financial backers or customers. An early trial was in 1898, in connection with a yacht race in Ireland, the Kingstown Regatta. The Dublin *Daily Express* arranged for Marconi and a reporter to be aboard a tug boat with a transmitter following the yachts and sending progress and final reports to shore faster than ever before. Yacht racing was far more popular then than now. A similar arrangement, sponsored by the New York *Herald*, was used at the America's Cup races at Sandy

Hook, New Jersey, in 1899. These were the first on-the-spot electronic reports of an event. There were to be many more.

Later in 1899, Marconi set up an experiment to transmit a signal from the vicinity of Dover, England, to Wimereux, France. Now he had proven that his methods could successfully span a distance of some 56 kilometers, over water.

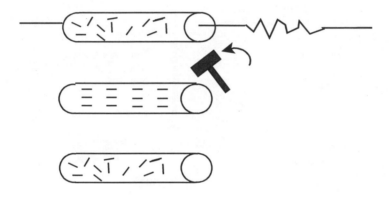

Figure 9.5. *Marconi's coherer, an early amplifier.* Small bits of nickel are randomly arranged in a container. In this stage, collectively, they are nonconductors of electricity. If an electric current is applied the pieces snap into alignment like soldiers coming to attention. In this arrangement they become a conductor and amplify the applied current. When the applied current stops, a hammer strikes the container randomizing the particles again.

Perhaps his best known demonstration was in 1901, transmitting from Poldhu, Cornwall, the westernmost part of Britain, to a receiver in St John's, Newfoundland, the easternmost part of North America. Again, it was a success, although lacking in the poetic content of Morse's first public message. Marconi sent the letter S in Morse code, simply • • •. But it was an intelligible message and it crossed the Atlantic Ocean without wires.

Meantime, he had set up a company in England, first called the Wireless Telegraph and Signal Company, then Marconi's Wireless Telegraph Company, and later the Marconi Company, Ltd. He applied for his first patent in 1896. Thereafter, there was a steady series of patents and patent law suits.

In all the early work there was little control over the frequency of the electromagnetic waves. That meant that, if there were several transmitters in a given region, they might all interfere with each other and no receiver would hear a clear signal. Marconi developed a means of tuning, both for transmitters and receivers, so they could select the frequency at which to operate. Then, if two stations in a region were using different frequencies, they would not interfere. Today, interference in broadcasting is a rarity because there are governmental controls over the use of frequencies and careful tuning is both possible and required. Marconi received patent number 7777 in England in 1900 for his tuning device and this became famous as the "four sevens." It was also famous for its court challenges.

Omitting the details, his claim was disputed, and as late as 1943, the United States Supreme Court ruled that Sir Oliver Lodge, Nikola Tesla, and John Stone had priority of invention and nullified Marconi's claim in the United States.[4] The American courts were involved because of use of the technology by American companies. By this time it did not much matter. Radio had developed into something quite different from what any of them anticipated.

Back to the early 1900s, radio began to draw the interest of some of the world's armies, navies, and merchant shipping lines. There were always seagoing accidents and on several occasions the wireless was able to summon help or inform families of passengers that all was well. By 1904 there were 68 Marconi land stations and 124 installations aboard ships. In 1907 transatlantic wireless telegraph service began. Marconi continued to grow and invent, sue and be sued over infringement claims. By 1912 the wireless was in common use aboard ocean liners. Common, yes, but it was not yet clear how it could best be used to assure safety.

The *Titanic* Disaster[5]

The story of the infamous *Titanic* has been told and retold. Its importance for us lies not in how it sank, how the passengers were rescued, or which ones were denied rescue, but in how communication facilities were used and misused.

The ship that was called unsinkable was the pride of Britain's White Star Line. Just before she was to leave for her first transatlantic voyage, there was a minor fire aboard. Afterward, it was found that the binoculars for use by lookouts in the crow's nest were missing. They were never replaced. Lookouts were expected to see (communicate visually with) icebergs before contact.

Once *Titanic* was underway from Britain to the United States, the director of the White Star Line, who was a passenger on board, wanted the ship to beat the time of a sister ship, *Olympic*, and asked that it maintain full speed. Ice warnings were heard on the morning of April 14, 1912. One of these was from *Californian* reporting that she was surrounded by ice and had stopped. She was either 19 or five miles from *Titanic's* eventual grave, depending on the source you believe. *Titanic* did not slow down. At 11:40 PM (local time) about 500 nautical miles off Cape Race, Newfoundland, she struck an iceberg. At 12:15 AM on the 15th, the captain ordered his radio operator to send out a distress signal. It is to some extent understandable that he waited that long, to verify the extent of the damage, and perhaps to avoid facing the reality that the unsinkable *Titanic* was sinking, especially since the line's director was among the passengers. But the delay proved fatal.

Several ships heard the call for help, at distances ranging up to 100 miles. *Titanic* cruised at 22 knots. While not the fastest ship afloat at the time, she was faster than most. A ship 100 miles away was going to need at least five hours to reach her. As it turns out, only *Carpathia*, from 58 miles away, was able to reach *Titanic* in time, in time that is, only to rescue those who had made it into a life boat, not those who could not get into a boat.

Californian had wireless but ten minutes before the distress signal was sent her operator had gone off duty, leaving no one monitoring the radio. *Californian* did see signal flares from *Titanic* but they were seen low to the horizon and their meaning was apparently not understood; nothing was done about them.

Titanic sank at around 2:30 AM on April 15. This was two hours and fifteen minutes after the distress call and two hours twenty-five minutes after *Californian*'s radio operator went off duty.

A bit of controversy, not really important to the main issue, arose over the role of David Sarnoff, then managing a receiving station for Marconi in the John Wanamaker department store in New York. Some shore stations did hear of the sinking and relayed the news around the country, the Wanamaker station among them. As a humanitarian service they reported the names of known survivors. Sarnoff was supposed to be on duty during the hours the store was open. The *Titanic* sinking occurred outside store hours. Nonetheless, he claimed, or it was claimed on his behalf, that he heard and relayed the messages. He couldn't have helped in the rescue. There was no air rescue service to be called, but the issue came up as Sarnoff went on to become head of the Radio Corporation of America and the vision of a heroic operator sticking by his radio during a crisis helped his reputation. To this day, no one seems certain what actually happened.[6]

Titanic Lessons

There are several communication lessons from all this.

First, the wireless showed its great potential for saving lives at sea, if properly used.

Second, if you are going to have a safety device aboard ship (the radio) then use it *all* the time, not just when it is convenient to do so. Calamities at sea do not happen on schedule. On the other hand, telegraphy required a skilled operator, not just someone to listen for voice messages. Ships' crews probably did not have enough operators to provide around-the-clock coverage. But when new technology comes along effective use of it requires adjustment to older operating procedures. Many people, to this day, have not learned that. And so, for example, we have automated telephone call centers that provide far worse service than what they replaced because they are unable to provide enough operators or sometimes enough equipment to handle all contingencies. *Titanic's* missing binoculars were a form of communication device, enabling objects to seem to be brought closer for viewing. Their absence was ignored. Would they have helped see the fatal iceberg at night? We do not know.

Third, an older, alternate means of communication—the flares —was available, used, and either ignored or misunderstood. There was a failure on the part of someone to see to it that all mariners were informed of such signals. Captain Lord of *Californian* did not have the curiosity or sense of responsibility to investigate an unfamiliar signal. In short, communication is not automatic. Senders and receivers must both understand the signaling systems in use. Lord was found negligent by a British Board of Trade inquiry but never faced charges in court.

Out of all this did come something of value. The North Atlantic Ice Patrol, an information service, was established to report ice conditions by radio to a central office. No ship has since been lost in the area due to a collision with an iceberg.[7]

Radar[8]

We interrupt the historical sequence at this point to introduce one of the most important uses of radio—radar. Radar is not a communications medium in the sense we having been using the term. We cannot send a wide range of messages by using it. Basically, we either do or do not receive reflected waves from some object, just as in radiotelegraphy we do or do not receive tones but are not concerned with tonal variations.

Radar is a means of sending out radio waves that, upon striking certain substances, such as the metal body of an airplane, bounce back toward their source. When the echo is detected the sender can tell that something is out there and can compute its distance by noting how long it took a wave to go out and come back. Its principal use in its early days was the detection of airplanes. The word *radar* is actually an acronym for **radio detection and ranging**.

Radar can be regarded as a means of communication between a machine, the radar set, and an inanimate target, the airplane. Work on radar during World War II gave birth to the idea of using these same high-frequency microwaves for more conventional communication, and later even for cooking food in microwave ovens.

Again, no single person can be said to have invented this system and made it practical. Early work was done independently in the 1920s and 1930s in England, Germany, Japan, Italy, and the United States when various investigators realized that radio waves being sent out could be detected if reflected back. But what use to make of this? Marconi was among the early experimenters. The Italian government realized the military potential and insisted that work be restricted to Italians only. Nothing came of this work. Marconi had other interests. Germany and Japan each produced a working system but neither foresaw an application for it, as radar was originally an entirely defensive tool and, in the 1930s, defense was not these countries' priority. So, their development work was limited.

It was in Great Britain where at least some people foresaw the threat of war in Europe and realized how vulnerable their island would be to modern bomb-carrying airplanes. They had suffered some bombing from airships in World War I and the general consensus was that nothing could stop the modern bomber aircraft.

Robert A. Watson-Watt was assigned by the Royal Air Force (RAF) scientific establishment to look into some way to stop or detect incoming planes. Originally it was hoped he could develop a "death ray" that would kill the pilots but this was quickly declared unfeasible. His assistant, Arnold F. Wilkins, had heard of a Post Office study indicating that some radio receivers being used for communication were picking up some form of radiation that

seemed to be reflected from passing airplanes. This led to speculation that there might be some way to make use of these reflected
waves. While this was interesting as a scientific idea, it was a time
of attempts at worldwide disarmament by the Western democracies
even while other countries were rearming. As a result, in Britain
there was a shortage of money for military research. More than
good science was necessary. There had to be both military and
political support. The principal military proponent of radar was
Hugh C. T. Dowding, soon to be an air chief marshall and head the
RAF Fighter Command. Dowding, Watson-Watt, and Wilkins get
the lion's share of credit. The fascinating story is told in detail by
David Fisher.[9] In short, a chain of radar sets was developed, each
able to detect the presence of incoming airplanes and their approximate altitude and direction from the detector. These were installed
just as Nazi armies were overrunning France. Hitler's next major
objective was England and the chosen method was to eliminate the
RAF as an effective force, then either force England to sue for
peace or launch an invasion of that country. The bombing began in
earnest in the summer of 1940, in what was called the Battle of
Britain. RAF bases and planes were the primary objective, but
industrial sites and the city of London were soon added to the
target lists. London was hit very hard.

These attacks came very close to meeting their objectives but,
as radar improved, the relatively few remaining RAF pilots and
planes began to turn the tide and destroyed so many German planes
that the invasion was put off. It was a close call and led to Winston
Churchill's famous tribute, "Never in the field of human conflict
was so much owed by so many to so few."

The Germans then concentrated on submarine warfare, hoping
to beat Britain by blocking food and munitions shipments into their
island. Again, they very nearly succeeded. But, as shipping losses
were reaching an intolerable level, radar sets had been re-engi-
neered to fit into airplanes with the result that U-boats could be
detected, then sunk from the air and, again, a tide was turned, this
time sealing the fate of the Nazi war machine.

Two other major applications of radar were developed early in
the war. One resulted from the fact that radar could see airplanes

but not tell which were friendly, which enemy. A device was put into British planes that detected an incoming radar signal and triggered a coded response. The operators on the ground saw not just the result of reflected waves but a distinct pattern that could have only come from an equipped, friendly plane. This was called Identification Friend or Foe (IFF), a form of which is still being used in air traffic control. The second development was called Ground Controlled Intercept (GCI). By watching how enemy planes approached and friendly interceptors rose to meet them, radar assisted controllers on the ground could direct the fighters to the right location and altitude for a successful attack, even in heavy clouds or the dark. Later, this idea was expanded to become Ground Controlled Approach (GCA) that enabled controllers to tell a pilot flying through fog how to approach an airport and, by providing information on appropriate heading and rate of descent, could guide the plane down to the level where the pilot could see the ground, perhaps by then only a few feet under the plane, and complete the landing under direct pilot control.

Today, radar is routinely found at all but the smallest airports to guide planes, in commercial airplanes to detect storms and other planes in flight, and in ships and even rather small boats to detect other craft at sea.

The basic idea is simple. Send out radio waves and provide a receiving antenna that can detect those waves upon being reflected back. This requires sending out a short burst of waves then waiting to see if anything comes back. After experimenting with waves of various lengths, the British settled on those in the microwave band, around 1000 mHz. If the outgoing signal were continuous, it would be almost impossible to detect the reflection, so the decision was to send a pulse, wait for an echo, then send another. Knowing how long it took to get the reflection easily yields the distance, as shown in figure 9.6. Originally outgoing signals went out from nondirectional antennas mounted on tall steel towers, much as we see today, used for radio or TV transmission (figure 9.7). To tell the direction to an incoming plane, it had to be detected by more than one side-by-side antenna. To tell how high up the plane was, it had to be seen by more than one antenna, mounted one above the other. Eventually, we learned to narrow the outgoing signal to a thin beam in order to tell from what direction the reflected signal is coming. That in turn meant that the antennas had to be able to rotate to sweep the thin beam over a wide arc. Altitude could then be measured by pointing the antennas up and down.

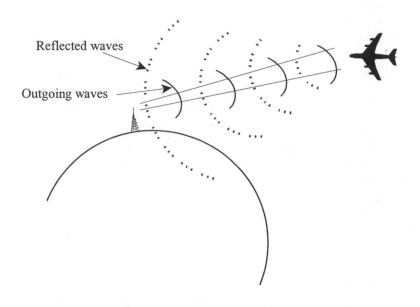

Figure 9.6. *Basic concept of radar.* A radio signal is sent from a ground-based antenna. When it strikes a passing airplane, some of the waves are reflected back. When the ground station receives the reflected signals it can tell how long it took for a given pulse to go out and come back. With that information it is a simple calculation to determine how far away the plane is.

If you pass a major airport today you will typically see several large spherical enclosures, usually white, that contain radar antennas, protected from the elements. If you look at a ferryboat in the harbor or one of the larger yachts in a marina you will see a smallish box on a mast, perhaps 15 cm. high and one or two meters wide, containing the modern radar set. Figure 9.8 shows a modern antenna used on a seagoing sailing ship. Figure 9.9 shows an antenna dating from about the 1950s having a parabolic reflector to form a narrow beam and an enclosed antenna used on a modern small ship.

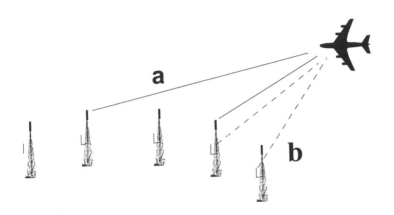

9.7. Early form of radar antennas. The original antennas, used during the Battle of Britain, were mounted on towers along the coastline. Since signals from any tower went in all directions, more than one station (solid lines) had to pick up the reflected signal in order to tell from what direction it was coming, based on the difference in time to receipt. Similarly, to measure the plane's altitude, it was necessary to send signals from two different heights (dashed lines) and again compare the time difference of incoming reflections. Sounds complicated but it worked.

Like wireless telegraphy used to transmit ice warnings, radar has become a major lifesaving device. The intense interest in microwave radio propagation led to its use in telecommunication shortly after the war.

Figure 9.8. *A modern radar antenna on an old-fashioned ship.* The radar antenna on this seagoing sailing ship is barely visible atop the mast on the right and shown enlarged in the inset. Photo courtesy Nautical Adventures.

Radio Broadcasting

Back, now, to the First World War era. With the commercial success of wireless a whole new group of inventors and entrepreneurs became involved in pushing the concept to its limits, both by improving wireless telegraphy or extending it to carry the human voice. Shortly after voice transmission became possible, broadcasting, rather than point-to-point or person-to-person communication, began.

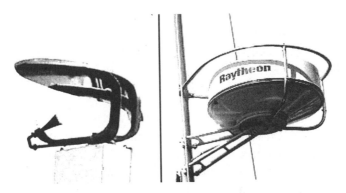

Figure 9.9. *More modern radar antennas.* On the left is an antenna used on a seagoing ship around the 1950s. Radio waves come from the small device on the left and are sent to the parabolic antenna which sends out a focused beam so there is no need for the array of antennas of old. The right-hand illustration is a modern antenna aboard a small coastal ship. It is about a meter in diameter, protected against the elements. Photos by C. Meadow.

The New Inventors and Prophets[10]

It would be hard even to list all the participants in law suits, let alone all those who made some substantive contribution. The following were the key players.

Thomas Alva Edison (U.S.A., 1847-1931) was associated in some way with just about every major electrical invention of his time and he was probably the last of the great inventors without much formal education in his field. He contributed to telegraphy and telephony and, while working toward an electric lightbulb, discovered a curious phenomenon in 1883 that came to be called the *Edison effect*. He had a glass tube approximately the shape of a modern light bulb. Air inside was evacuated, leaving a near vacuum within. Electric current coming from a wire filament (figure 9.10) would jump a gap over to a small metal plate. In other words, the current continued to flow, even without a conductor between these two metal elements. It didn't much help his electric lamp work, so nothing more was done with it at the time, but he did patent the device.

Sir John Ambrose Fleming (England, 1849-1945), once a consultant to Edison, used the Edison effect in 1904 to produce direct current (DC) from alternating current (AC) input. DC was needed to power a radio or telephone speaker.

Lee de Forest (U.S.A., 1873-1961) was a Yale PhD who developed the first modern-style vacuum tube in 1905. He discovered that he could enhance the Edison effect by adding a third element within the tube, another wire running between the filament and the plate, shown at right in figure 9.10.

By putting an electric charge on this third wire he could increase the amplification of the incoming current and could vary the rate of amplification by varying the charge on the third wire. He was obsessed with becoming rich and famous and gave no credit to Edison or Fleming for his work. Sometimes called a *triode*, de Forest called his tube the *audion* and initially considered it only as an aid to wireless telegraphy.

Edwin Howard Armstrong (U.S.A., 1890-1954) became interested in wireless as a teenager and was later educated in engineering at Columbia University. In 1912 he worked out a method of increasing the amplification of the de Forest audion. He is best known for developing a circuit that could receive and convert very high-frequency transmissions for use by the receivers of the time and, later, for inventing frequency modulation (FM) for use with radio.

During the First World War, radio was used by both sides for military communication and, of course, transmissions were intercepted by both sides. The Germans found a way to transmit at very high frequencies which the Allies, to their great consternation, could not intercept. Armstrong was then an officer in the army and was asked to look into the problem. He devised a circuit, called a *superheterodyne*, that, leaving out the physics, would convert the high frequencies to lower ones that could be easily received and transformed to DC. It was also a tuner, allowing for selection of

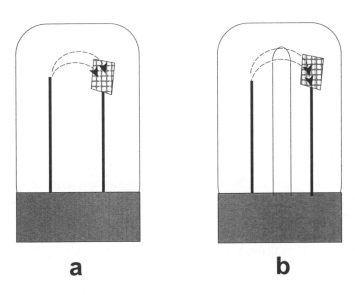

Figure 9.10. *The Edison Effect and the de Forest Audion.* In a glass tube from which all air has been evacuated are two metal electrodes, **a,** not touching. As a current is applied to one, a spark jumps the gap between them, completing a circuit. This is the *Edison effect.* Fleming noted that, if the input current is alternating, the output is direct, more useful in powering radio speakers. Later, de Forest added a third element to the Edison tube, **b,** placing a wire grid between the two electrodes, to which a variable current is applied. The amount of current controls the amount of amplification of the basic current.

which frequencies were to be received at a given time. The army was glad to use it and did not insist on patent rights. After the war Armstrong "went commercial" and his device was used eventually in all radio receivers.

David Sarnoff was born in Russia in 1891, immigrated to the United States as a child in 1900, and died in 1971. He began his career selling newspapers and soon dreamed of becoming a reporter. He went to the building housing the New York *Herald* to ask for a job but mistakenly entered the office of the Commercial

Cable Company, also owned by *Herald* boss James G. Bennett. This was a telegraph service. They had an opening for a messenger and offered him the job, and so an illustrious career in telecommunications began. Between deliveries he watched the telegraph operators at work, bought his own key, and taught himself to use it. He soon ran into a problem. He was fired because he would not work on Jewish holidays, but quickly landed another job as office boy with American Marconi, an association that was to last the rest of his career through corporate reorganizations. He gradually worked up to an operator's job, then station manager, and then station manager at the prestigious Wanamaker store's office. Eventually, he became assistant chief engineer of the company.

In 1913 Sarnoff met Edwin Armstrong and the two quickly became friends. Armstrong's receiver was very advanced but Marconi was indifferent so Armstrong brought it to AT&T, giving that company a potential edge in competition with Marconi. In 1914 Sarnoff experimented with transmitting music and he began to think about the concept of broadcasting, transmitting to a wide audience all at once. He was not alone, several others had the same idea. In order to make broadcasting practical, though, there had to be a large number of radio receivers in use. Sarnoff called these *radio music boxes*. They first had to be designed, manufactured, and sold. But as these thoughts were brewing in his mind, World War I was brewing in Europe. It was no time to propose a whole new industry. However, just recognizing the need for manufacturing a large number of new consumer devices meant that Sarnoff-Marconi clearly had a lead on all competition.

In 1917 Sarnoff was made commercial manager of American Marconi Co. The war accelerated wireless progress, Armstrong's work being just a part, though a major one, of this development. Radio was used for communication by both sides, on land and on the sea. During World War II Sarnoff served in the U.S. Army, planning the radio communications for the Allied invasion of Normandy and achieving the rank of brigadier general.

Reginald Aubrey Fessenden (Canada, 1866-1932) once worked for Thomas Edison. As early as 1900 he conceived of a wave generator that could produce waves of 100,000 Hz, enough bandwidth to carry all the needed variations for voice and music transmission. He took his idea to the General Electric Company in 1900, but they were unable to produce such a device until 1906. Using it, Fessenden became the first person to broadcast speech by wireless. However, his transmitter was expensive and few people had receivers, so it remained but an interesting artifact for some time. He also developed the heterodyne circuit, which Armstrong later made into the superheterodyne. A Dane, Valdemar Poulsen, also independently managed to transmit voices, but his equipment was never made practical.

Spanning a Nation[11]

The First World War had brought technological progress and widened the interest in and use of wireless. Armstrong had his superheterodyne and the Marconi Company, like all others, was looking for new markets. The world was almost ready for broadcasting.

Both Marconi's U.S. and British companies needed some hardware from General Electric Company in order to further their wireless efforts. The U.S. government was reluctant to see too much foreign domination in the new but increasingly important radio industry. Proposals were being made for government control over the industry. To counter this, in 1919 representatives of the Navy Department and Owen D. Young, a vice president of General Electric, proposed that GE and American Marconi set up a new, jointly owned company, to be called Radio Corporation of America. Westinghouse and AT&T also had shares. Edward J. Nally, Marconi's president, was to be its president. David Sarnoff was to be its commercial manager. Nally asked Sarnoff to assess the new company's needs and business potential. He, of course, proposed his radio music boxes as a new line of business.

At about the same time, GE's major competitor, Westinghouse Electric, set up radio station KDKA in Pittsburgh and broadcast presidential election returns in 1920. Other stations followed in St.

Louis, Newark (New Jersey), Springfield (Massachusetts), and Chicago. The race was on. Both Westinghouse and RCA found broadcasting catching on with the public. Other companies, with names no longer recognized, such as Atwater Kent, Crosley, Philco, and Stromberg-Carlson, began selling receiving sets. In the two decades from the start of broadcasting in 1920 to 1940 growth was impressive, with an early spurt in the 1920s, then a leveling off as first the depression and then war inhibited growth. After World War II, growth was explosive, beginning to slow down only as television and then FM came into use. The industry still shows growth in spite of its competition. Figure 9.11 shows the number of radio stations in the United States from the beginnings of broadcasting to the 1990s.

Key Events

In 1921 a speech by then Secretary of Commerce Herbert Hoover was broadcast. Speeches by secretaries of commerce are not normally landmark events but this was the first of its kind and led Hoover to decide in 1928 that, if he were nominated for president, he would campaign "mostly on radio and through the motion pictures."[12] This *was* significant.

Ironically, it was Hoover's eventual adversary, Franklin D. Roosevelt who, upon inauguration as president in 1933, initiated a series of radio talks he called *fireside chats*. With these talks he became "one voice from Washington [that] was able to unite the nation as none other."[13] A close associate, Samuel Rosenman, described his impact on the radio audience as, "When his voice came over the radio, it was as though he were right [there] . . . discussing their personal problems with them — the cattle or crops of the farmer, the red ink of the shopkeeper, the loans of the banker, the wages of each worker."[14] Roosevelt was perhaps the first great master of the new medium. Nothing like this had happened before.

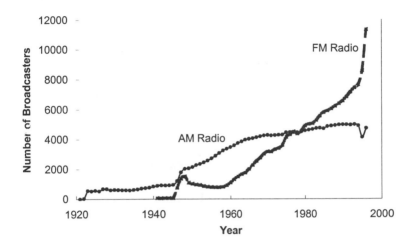

Figure 9.11. *The number of radio broadcasting stations in the United States, 1921-1996.* Broadcasting began in the United States, using AM, in 1921 and the number of commercial stations has grown fairly steadily ever since. FM was developed in the early 1940s, grew slowly because of the war, then spurted upward, leveling off a bit, then beginning a steep increase still going on. The two media have learned to specialize in different kinds of broadcasts, AM mainly news and talk, FM music. Radio is far from dead.

An unexpected event in 1937 gave us our first live voice broadcast of a dramatic event. The German passenger-carrying airship *Hindenburg* was landing at Lakehurst, New Jersey, when it burst into flames. Herbert Morrison of radio station WLS in Chicago was covering the event. His emotion-laden reporting of the fire and loss of life brought a vividness never before possible in a news story and was called by the Chicago *Tribune*, "the most gripping thing they ever heard."[15]

Just three years later, the Second World War was underway. Edward R. Murrow was reporting by radio from London for the

Columbia Broadcasting System. His calm, eyewitness accounts of the Battle of Britain, with its relentless bombing of London, brought an unaccustomed vividness to his listeners. He set a new standard for broadcast journalism. The following quotation may not sound dramatic to modern readers, but delivered in Murrow's calm, somber style, at a time when listeners were unused to such reports, it seemed to bring the hearer to the real thing:

> In the course of the last twenty minutes there's been considerable action up here. . . . Just straight-away in front of me the search-lights are working. I can see one or two bursts of anti-aircraft fire far in the distance. Just on the roof across the way I can see a man wearing a tin hat, a pair of powerful night glasses to his eyes, scanning the sky. . . . There is a building with two windows gone.[16]

What this man was telling us was what he was seeing, not last week, but *at the moment he was speaking*, and what he was seeing was something hardly anyone else had ever seen, the heavy bomb-ing of a major city. And yet, he could not literally make the audi-ence see what he was seeing; he had to get the audience to visual-ize what he was describing.

During the twenties, thirties, and forties entertainment also grew on radio. The family would gather around the radio to hear Amos 'n' Andy, Jack Benny, Bob Hope, the Lux Radio Theater, or the Bell Telephone Hour. Sports broadcasting began. For a while, it was too expensive to send announcers and engineers to out-of-town games, so a description was sent from the game site to distant stations in highly condensed form, by telegraph. The local an-nouncer "recreated" the game as if he were actually seeing it.

President Ronald Reagan had been such a broadcaster at one time and he told the story of once having lost his incoming signal in the middle of a baseball game in the early 1930s. Dizzy Dean was pitching. As one pitch left his hand the wire went dead. Reagan, now unable to see or hear the game, had to make up and describe action for his audience for nearly seven minutes. He claimed the batter was fouling one pitch after another. Finally, the line was restored and he learned the batter had fouled out on the first pitch.[17] While an amusing anecdote it indicates the reality of live broadcasting in those days. There is something of a parallel with the *Titanic* incident. *Californian* did not hear *Titanic's* signal because the radio operator went off duty. In both cases there was no provision for assuring that coverage would be available all the time, something broadcast hearers came to expect. They don't like dead air at any time and in emergency situations it can be deadly.

A final example of radio's power was the 1938 dramatization of H. G. Wells' story about an invasion from Mars, called *War of the Worlds*, produced by Orson Welles. This version was set in New Jersey and, in spite of repeated disclaimers that it was just a play, it created widespread panic among the listening audience, who thought they were hearing the news.[18] Vividness could be carried too far.

Television Arrives

The high point of radio broadcasting probably came during World War II because there was so much important news, the medium had become so wide spread, and other forms of entertainment were at a minimum. But just after the war television became a serious competitor. I saw my first TV broadcast in 1940, during the presidential election campaign of that year. It was quite dull, anything but vivid, even for a ten-year-old. All I saw was a rather grainy picture of a talking head. It looked like a badly made motion picture. But, fast as radio had grown, TV grew faster and soon began to take away entertainment and sports programs and then to develop its own news personalities, Walter Cronkhite, the team of Chet Huntley and David Brinkley, and Ed Murrow (again).

In 1954 Senator Joseph McCarthy, who had made his reputation by claiming to have uncovered Communists in key positions in the U. S. government, got into a conflict with the U.S. Army around this same issue. A Senate committee investigated the dispute and held hearings that were televised. These drew huge audiences. Vividness again. Marshall McLuhan pointed out that "McCarthy lasted . . . a very short time when he switched to TV."[19] Previously, he had used all the news media but on his own terms. Now he was on someone else's turf and he did not look good. Politics aside, he wasn't a TV personality. His reputation did not survive and, in fact, he did not survive physically for long after this.

Television took away audiences from radio for entertainment, sports, and news. But it did not kill radio. Where once there would have been one receiver in the home, probably the central piece of furniture in the living room, now there might be one in every room as well as in every car or truck. By the 1950s radios were also showing up in purses and pockets. Programming changed to being largely music and news. Sports events were still important, but now as an adjunct to TV coverage. Today, we listen to radio at home, at work, or while walking, jogging, or driving. It has survived by adaptation.

Some nostalgic notes: in 1985 General Electric, co-originator of the initiative to create RCA, bought the company. The name is still seen on some products but it is no longer the power it once was. The name Marconi no longer appears as part of any major United States corporation. General Electric, one of the sires of RCA out of Marconi, is now the survivor of both.

Social Effects of Radio Broadcasting[20, 21]

What radio offered that no other communications medium had ever offered before was instant, simultaneous, and vivid communication to a large number of people. Had you known Ed Murrow personally and been able to afford the cost of transatlantic radiotelephone in 1940, you could have heard his views on the war. But broadcast radio brought those views to a huge audience, all at once, and like President Roosevelt, he sounded like he was talking directly to each member of his audience. When an event was going on live, be it an entertainment show, baseball game, or political speech, the audience all heard it together *while it was happening*. That accounts for frequent use of the word *vivid* in this chapter.

The effect of this medium was to create a community of listeners. We all shared the same experience at the same time. President Roosevelt used this for good, to help encourage and energize a nation numbed by the economic depression of the 1930s. Mayor Fiorello LaGuardia of New York made friends by reading the comics over the radio to children (many of whom were doubtless quite grown) during a strike of newspaper deliverers. Demagogues used radio, too: Huey Long in Louisiana and Father Coughlin in Detroit, in the 1930s. This was not quite McLuhan's global village but it was a community and it was something we never had before.

Newer Technologies

One reason that radio survived competition from other media is that it is so versatile. It is a medium in at least two of the senses defined in chapter 2. It is a means of transmission and it is a group of organizations that create and distribute content. In the transmission sense, radio keeps on growing. In the content sense, it has lost much of its influence in the political and entertainment worlds but retained it in popular music. Television, as we shall discuss at greater length in the next chapter, is not a new means of transmission; it is a new way of using an existing means. TV transmissions go by radio, telephone wire, or cable. Telephone, which can be used to carry radio and TV programs to remote stations for over the air transmission at the new locality, sometimes sends its messages by radio. Here are a few developments:

FM Transmission[22]

The same Edwin Armstrong who invented the superheterodyne circuit, developed frequency modulation (FM) transmission. FM is a technique that encodes sound differently from the older amplitude modulation (AM). It uses higher frequencies and is able to transmit a wider range of sound frequencies, giving higher fidelity of sound. That makes it preferable for music which has become much more important to radio than it was in the first two decades. These high-frequency transmissions follow a line of sight. The lower frequencies of AM can reach farther but are subject to more interference. Interference comes from natural sources, such as electrical storms, and from other stations whose signals have reached far from their own broadcast region. With FM, interference from other stations is unlikely.

FM broadcasting began in 1940 and today is fast becoming the transmission mode of choice for broadcasting. Stations limited to AM are losing audience, except for news and sports which do not need the signal clarity of music broadcasts.

Microwave Transmission

Remember the early days of telegraphy and telephony, when wires had to be strung, sometimes over long distances and difficult terrain? What if we could send telephone messages across a river by building a steel framework tower on each side and transmitting radio signals across the water? It would be much cheaper and easier to maintain. This is exactly what has been done. Figure 9.12 shows some microwave antennas on the roof of a bank building in a large city. Banks, with their many branches, need to deal with other banks and large customers, exchanging digital information, and are heavy users of modern telecommunications.

Another use of microwave is in connection with communication satellites. The early ones were simply reflectors but they now contain repeaters and amplifiers of signals. The transmissions up to and down from it are by microwave.

Figure 9.12. *The new downtown skyline.* Antennas rise like trees from the tops of modern office buildings, in this case a large bank in downtown Toronto. These represent a variety of uses. Banks are major users of telecommunications. Photo by C. Meadow.

New Broadcasting and Receiving Methods

Radio just will not die. Today it is possible to get radio programs, live or recorded, over the Internet. Even newer is the notion of transmitting radio from communication satellites, much like modern-day satellite television. The advantage is that for something like a hundred selected stations, a listener on the move can keep listening to the same station wherever he or she may be.

There is also a new kind of receiver, powered by a hand crank rather than a battery or household electric current. The user cranks for a minute or so and gets around twenty minutes of listening time. The point is that travelers or those living in remote places can make use of radio without any source of electricity other than the crank.

Radiotelephone[23]

It is relatively easy to connect a radio and a telephone since both encode sound electrically. The earliest transatlantic telephone calls were not made through telephone cables, but by radiotelephone, that is, by just such a connection between the telephone network and a radio station able to transmit overseas. More recently, we have had *portable telephones*, the handsets of which are radios. As you walk around your house or out to the pool, you communicate with your base station by radio, it converts the signals it receives into those appropriate for telephone and connects to the wire telephone network. The range of such radios is normally limited to perhaps a hundred meters.

Then came *cellular telephone* in which, again, what looks like a small telephone handset is actually a radio. It has a range on the ground of hundreds of meters to several kilometers. While walking, driving, or sitting in a restaurant you talk by radio to a local radio station which relays messages to and from a central station, in turn connected to the wire telephone system. It is much like the portable phone, but with longer range, except that as you move from the domain of one cell to another, the phone system senses your movement. It does this because it hears your conversation at more than one cell and as reception grows faint at one cell and stronger at another, your call is handed off from cell to cell as you travel. The net effect is like having a very long range portable phone with a fast servant running after you to reattach the base unit whenever you move out of its range.

Next came *personal communications systems* (PCS) that are much like cell phones but use higher frequencies and can be connected to a satellite, giving a much greater range of coverage and greater bandwidth, allowing for more simultaneous calls. More of this in chapter 11.

An advantage to the cell phone and PCS besides their portability is the lack of a need for the telephone wire infrastructure. In parts of the world that lack such facilities, it is far easier to build a PCS system than a wire-based system, and operating costs for users are about the same.

One of the basic inventions used in these modern radiotelephone systems was made by the movie actress Hedy Lamarr and musician George Antheils. Ms Lamarr (née Hedwig Maria Kiesler of Vienna, 1913-2000) and Mr. Antheils were both escapees from Nazi-dominated Europe just before World War II and they both wanted to contribute to the war effort. It was her idea, which he helped to implement, to rapidly switch the frequency at which a radio signal was sent, adding some random sounds to it. These changes to the basic signal were governed by a key that had to be available to both sender and receiver. This technique would make interception and decoding difficult to impossible without the key. The inventors intended the system to be used for radio-guided torpedoes and toward that end they gave their patent rights to the U.S. government. However, at that time it required simultaneous use of a roll of paper tape at either end to govern the changing of frequency and adding or removing the extra sounds. It was never used during the war. The inventors felt that because they likened the tape to a player piano roll it completely turned off conservative navy brass. But, by the 1960s, digital equipment obviated the need for the tape, and the method was successfully used in military applications. Then it became basic to cellular and PCS systems.

Transistors and Integrated Circuits[24]

Vacuum tubes, as noted previously, were typically about the size of a 40 or 60 watt lightbulb. They had limited lifetimes and generated considerable heat. Larger machines, such as the early computers, required huge air conditioners to counter the heat. A transistor, originally, was about the size of three stacked dimes, did not generate heat, and lasted a very long time. The replacement of tubes by transistors made electronic machinery that was considerably smaller and lighter, lasted longer, produced less heat, and cost less. All this made it practical to make machines with more elements, giving computers, for example, far more logical power.

The transistor plays essentially the same role as the triode vacuum tube (the generic name for de Forest's audion). It is a solid, not something filled with gas, hence the term *solid state* to differentiate the new components from the older ones. It can amplify current, or serve as a switch or an oscillator (generating current that increases and decreases regularly). It was invented in 1947 by three scientists at Bell Laboratories, John Bardeen, Walter

H. Brattain, and William B. Shockley. So important was their work that they shared the Nobel Prize in physics in 1956. The first computer I worked on in 1954, called the UNIVAC, used vacuum tubes and took up the better part of a room, with another room for its air conditioner. It was far slower and with far less memory than any of today's laptop computers.

Miniaturization did not stop with the transistor. The next step was to develop *integrated circuits,* done in 1958. This is a piece of silicon in which are embedded tiny transistors. It became possible to put all the circuitry of a computer's main processor in a single "chip" about an inch square containing *millions* of modern-day transistors. And again, these are sturdier, cheaper, and produce less heat. These are what give us our present-day powerful desk top computers and telephones that fit in a shirt pocket. The inventors of integrated circuits were Jack Kilby, of Texas Instruments, who also won a Nobel Prize for his work in 2000, and Robert Noyce. The latter was once a coworker on the Bell transistor team, then joined Fairchild Semiconductor, and later became a cofounder of Intel. He developed the method of actually producing integrated circuits. Noyce's work was described as "the key invention of the twentieth century because it impacts everything, from education to products . . . and the way we deal with each other in society at large."[25] The modern version of every electronic device discussed here, radio, television, telephone, and all their variants, depend on solid state, integrated circuitry.

To give an idea of the directions this miniaturization is going, some recent research at the University of California at Berkeley proposes "smart dust," sensors and computers so small they can float on the air like dust particles, observing what is going on below, whether a battle or weather phenomena, and reporting to a ground station.[26]

Radio has come a long way since Marconi and shows no signs of disappearing from our world. On the contrary, wireless portable telephones are becoming extremely popular and communications satellites used for telephone and television employ microwave radio. The technology is steadily improving and this leads to change in the nature and content of the messages carried.

Notes

1. Baker, *History*; Dunlap, *Marconi*; Jolly, *Marconi*; Lewis, *Empire*; Crowley and Heyer, "Radio Days."
2. Lewis, *Empire*, describes the work of de Forest, Armstrong, and Sarnoff. Baker, *History*, briefly describes the early work of Faraday, Maxwell, and Hertz in nontechnical terms. See also Rowlands and Wilson, *Oliver Lodge;* Dunlap, *Marconi*; and Jolly, *Marconi*.
3. Baker, *History*; Lewis, *Empire*.
4. The citation for the court's decision is 320 U.S. 1, 1943. Brief, nonlegalistic accounts are found in Aitken, *Syntony*, 167-68, 258; and Rowlands and Wilson, *Oliver Lodge*, 168, 199.
5. Baker, *History*; Gardiner and van der Vat, *Riddle*; Kern, "Wireless World."
6. Lewis, *Empire*, 105-107.
7. Jolly, *Marconi*, 186.
8. Fisher, *Race*.
9. Fisher, *Race*
10. Brown, *Manipulating*; Crowley and Heyer, "Radio Days"; Lewis, *Empire*; Raby, *Radio's*.
11. Brown, *Manipulating*; Crowley and Heyer, "Radio Days"; Lewis, *Empire*; Raby, *Radio's*.
12. Lewis, *Empire,* 182.
13. Lewis, *Empire,* 238.
14. Rosenman, *Working*, 249.
15. Brown, *Manipulating*, 141-144.
16. Brown, *Manipulating*, 185.
17. Reagan and Hubler, *Early Life,* 66-67.
18. Brown, *Manipulating,* 201-253.
19. McLuhan "Understanding Radio," 252.
20. Fisher & Fisher, *Tube*.
21. Crowley and Heyer, "Radio Days."
22. Erickson, *Armstrong's Fight*; Ladd, *Radio Waves*.
23. Harte, *Cellular and PCS*.

24. Bernstein, *Three Degrees*; "Electronics: Invention of Transistors,"
    "Transistor, Integrated Circuit," *Encyclopaedia Britannica*; Reid,
    Chip; Solyman, *Getting the Message*, 192-203.
25. Quoted in Richards, "In Noyce's Passing."
26. Kupfer, "Spies in the skies."

10

Television[1]

Introduction

Sometimes it seems there must be a human gene that makes us want to communicate by creating, sending, and receiving pictures. Gestures are a kind of picture of brief duration. Writing is also pictorial, although long lasting. Pictures are the oldest form of recorded communication we have found, but initially their use required the viewer to come to the cave to see the picture on the wall. Later, the message could be delivered as clay tablets or shapes of various kinds. As we moved into electrical and electronic transmission, we had the early work of Alexander Bain who invented a form of facsimile in 1843, long before it had any practical use. But the urge to transmit pictures remained; not only to send stills but live pictures of moving people and objects, pictures of events while they were happening

What is the attractiveness of images? Guy Vanderhaeghe put it, in the context of viewing moving images that they

> burn themselves in the mind. . . . A book invites argument, invites reconsideration, invites thought. A moving picture is beyond thought. Like feeling, it simply is. The principle of a book is persuasion; the principle of a movie is revelation.[2]

Beginnings[3]

As we said in chapter 9, the electronic age began when Michael Faraday discovered magnetic induction, Maxwell recognized electromagnetic waves, and Hertz built the first radio-wave transmitter and receiver. Then there was the Edison effect, not immediately recognized for what it was. Two of the basic television inventions were an early form of camera and the cathode-ray tube that is used to display an image to the viewer. Efforts to develop a working system went on in England, France, Russia, Germany, and the United States.

To send a written message by telegraph, whether wireless or not, the operator would scan the text and convert each letter to a

Figure 10.1. *The scanning of a graphic image.* A light beam scans from the top, left to right, then down to the next line. The intensity of light at any small area is averaged and converted to intensity of current. Shown here in exaggerated scale is a scan of the top two lines of an image.

code, then transmit that code using a telegraph key. You cannot do this with pictures. There is no code for every possible image in a picture. Instead, it is scanned line by line as in figure 10.1.

During the scan an electric current or a sequence of digital values is generated that corresponds to how light or dark the image is at any point as shown in figure 10.2. Let's deal with black and white first. The degree of darkness is called the *gray scale*. We might recognize shades of gray from pure white to pure black. Figure 10.3 shows variations in an image using, successively, 256, 4 and 2 shades or levels of gray. Today computers and television use far more shades, capable of showing very fine degrees of shade variation.

Figure 10.2. *Transmitting the graphic image.* The intensity of reflected light is converted to an analogous electric current which can be sent over a telephone line. The dotted lines here indicate the approximate level of reflected light at any point, a higher line indicating more white space, hence more reflected light. The solid lines simply indicate the scan lines, again in exaggerated scale. At the receiving end this is reconverted to light, either electronically or as marks on a paper.

Most people are familiar with the dots that make up a photo-graph printed in a newspaper, shown in enlarged form in figure 10.4. What from a distance looks like a continuous gradation of gray shades is actually a large number of tiny, discrete spots, called *picture elements* or *pixels*, each one a particular shade of gray. In television and facsimile, a whole line of the image is scanned and the gray scale values are transmitted continuously as the scanner moves from left to right. Note that in this form of representation of information, used in television and facsimile, the camera or, in modern times the computer, does not know it is scanning the letter A. It does not interpret an image, it merely copies what it sees. The camera only knows which parts of the scene it is scanning are light and which dark. In a telegraph or in a computer, the A is explicitly represented by a code for that letter and that alone.

Figure 10.3. *Variation in the number of shades of gray in an image.* An original picture (left) is reproduced using 256 levels of gray. The center image uses four, the right image two levels. The full 256 shades may not be detectable in this reproduction but the difference between it and the four level version is clear. The two-level image loses detail but is still barely recognizable. Photo by C. Meadow.

Interest in what became television began even before Marconi's work on radio, but progress was much slower and enormously more expensive. The very word *television* caused some upset when it was first used. The *Manchester Guardian* editorialized in 1935 that, "The word is half Greek and half Latin. No good will come of it."[4] This probably presaged the generally negative feelings of literate people toward the medium to this day. By *literate* I mean those devoted to the printed word. More on this later.

When work started seriously on what was to become television, there were many individuals and groups involved. They usually did not communicate much among themselves until the 1930s when the approach of commercialization spurred interest in mergers and acquisitions, and the British Broadcasting Corporation even set up a formal competition to decide which system they would use. During the Cold War, the United States and the Soviet Union seemed to like trading claims for inventions. Americans were probably more amused than insulted when Russians claimed television as one of theirs, but this was not totally baseless.

Figure 10.4. *Enlarged pixels*. This is a copy of a photograph as printed in newspaper, consisting of a large number of pixels. The farther away the viewer stands, the more the eye merges the separate pixels into what appears as continuous tones. Close up you can clearly see the individual pixels. Photo courtesy University of Toronto *Bulletin*.

The inventors did build on each other's fundamental inventions, some not directly tied to TV. We had the Edison effect and the audion which led to practical electron tubes that converted alternating current to direct and amplified the current at the same time. The Armstrong superheterodyne wound up in all radios and the early TV sets made use of existing radio receivers. Karl Ferdinand Braun (Germany, 1850-1918) invented the cathode-ray tube (CRT) in 1897 and this became the basis for television displays, to this day, although there is now some newer technology that uses a flat screen, taking up less room.

A. A. Campbell Swinton published a paper in the British scientific journal *Nature* in 1908 in which he proposed a design for an all-electronic television. But he never built one. In fact, in 1920 he indicated that television probably had no future commercially but he was still the first to propose essentially what became television.

Because there were so many people involved in experimenting with TV, we'll pause in the tale of what was done to give some background on how it works in general. This description probably does not fit any one design but describes the essence of most of them. Then we won't have to go into detail about each person's work. We'll also introduce the transmission of still images, the process now known as facsimile.

Technicalities

There are three major components of an image transmission system: a *camera* or *scanner* that "sees" a scene and converts it to electrical current, whether analog or digital, a *transmission system* to carry the image from the camera to the viewer, and the *display* that re-creates the image for the viewer. Facsimile is a less complex transmission system than television.

Facsimile[5]

A *facsimile* or *fax* scans an image line by line. The image may be text or graphics, but the fax machine is not aware of which. The black and white areas it finds as it scans are converted into analogous electric current as in figure 10.2, then sent through a telephone line. (There are now color faxes but they are still rare. We will concentrate on black and white.) At the receiving end, the procedure is reversed and the current from the telephone line becomes black and white areas on a paper, thus reconstructing the image.

Telephone is not the only means of transmission, and it was not even the first, but it was through telephone that fax caught on as an important, widely used medium. The idea was first developed in 1843 by Alexander Bain, then in Scotland, as noted earlier. Beginning in 1922 a form of facsimile was used to transmit photographs by radio. Called *wirephoto*, it was first used by the London *Daily Mail* and its inventor was the Canadian William Stephenson, later to become one of the leaders of British intelligence during World War II. The wirephotos were particularly popular during that period as a means of bringing "hot" war news to print media. The procedure was essentially as described for fax except that radio, rather than telephone, was the transmission medium.

In the late 1970s to early 1980s fax began to catch on in a serious way as an office machine. By now it is found in almost as many businesses as the telephone. It is also found in many private homes, such usage boosted by machines that are a combination of printer for a personal computer, photocopier, scanner, and fax. The output of a modern fax is indistinguishable from that of a photocopier because the scanning and print mechanisms are identical.

Fax enables the sending of documents that may or may not have graphics, and have not been digitized. It is even used to send legal documents, with the image of a signature included, now often accepted as a legal signature.

The Television Camera[6]

We cannot transmit an entire image all at once. It has to be decom-
posed into smaller elements and sent piecemeal. When the human
optical system does see an image, it persists in our brain for about
1/30 of a second. To give the appearance of continuity another
image must be displayed before that 1/30 second has expired. In
motion picture projection we are actually shown a series of still
pictures: one is projected briefly then the shutter on the projector
is closed, so momentarily nothing is projected, then another frame
is moved into place and the shutter opened. We see the slight
difference between successive still pictures as continuous motion.
Early projectors did not work so quickly, making the action appear
to flicker, hence the still-used nickname for movies, *flicks*. Television, then, has 1/30 of a second in which to scan an image, transmit
it, and display it. If it took longer, the image could be blurred if
anything moved during the scan. If it meets the 1/30 second
deadline before starting on the next image, our eyes see the action
as continuous. (Actually the TV makes two scans and displays
alternate lines in each scan. Each of the two partial scans must be
done in 1/60 of a second.)

There were two main challenges in television camera design,
how to scan the image and how to convert it into electric current.
One way to scan, developed by Paul Nipkow, is to have the camera
look at the scene through a moving peephole. Move the hole from
one side of the scene to the other, then switch to a lower peephole
for the next line. This is how the first successful camera worked.
It was mechanical, not electronic. It used a disk into which were
cut a series of small holes, as in figure 10.5. As the disk rotated,
anyone or anything looking through any one hole as it moved
through an arc would have a view like panning with a telescope,
that has a very small angle of view. When the end of one line is
reached viewing switches to a lower peephole. The scan is not
exactly horizontal, it's an arc of a circle, but a close approximation
to a line. Thus, the scene becomes a series of lines that are transmitted sequentially.

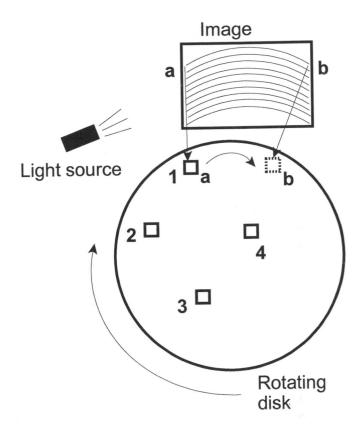

Figure 10.5. *The Nipkow disk.* The first television camera made use of a device like this to create the effect of linear scanning of the image. As the disk rotates, reflected light comes from the image through the top hole which moves horizontally across the image area. When the end of the line is reached, the camera uses the next lower hole, which views a line slightly lower in the image area. When hole **1** reaches position **b**, hole 2 is just under the original position of **1** (at **a**). Light coming through the holes is projected onto a screen. Actually, the projection is in the shape of an arc of a circle, but the curvature is not much noticed. Invented in 1884, this was used in the first television broadcasts in England in 1929. The drawing is not to scale.

The beam of light that came through the hole was projected onto a surface coated with some substance that emitted electrons when hit with light; the more light, the more electrons. This is called the *photoelectric effect*. That was the way to convert gray scale into electric current. A pure white spot in the image knocks loose a lot of electrons, a pure black spot, none. A gray spot is something in between. What material? The first one used was the element selenium, whose light sensitive or photoelectric properties were discovered in 1873 by Joseph May and Willoughby Smith, who worked for a British telegraph maintenance company. Selenium wasn't great in television because the current it produced was very weak, and it was slow to react, but at least it *did* produce a current. An amplifier was needed and this was initially the de Forest audion. Later, other, better photoelectric substances were used.

Whether the scene being scanned is still or moving doesn't make much difference with modern equipment. It has no trouble getting what amounts to a snapshot, what you might get with a photographic camera at 1/30 of a second exposure.

Transmission

Now we have a scene and we have it converted into electric current. The information can be transmitted by radio, just as the small current from a microphone is sent by telephone or radio. The radio receiver converts the electromagnetic waves back into electric current. John L. Baird, one of the major early developers, used to be able to tell something about the picture being transmitted from the sound coming from his radio receiver.[7]

Display

At the receiving end we need to convert the electric current into a visible picture. What is needed is the opposite of the photoelectric process, something that would glow or produce light at a spot if bombarded by electrons and glow with intensity proportional to the number of electrons hitting the spot. Here, there are three problems: how to produce a narrow beam of electrons, how to control its movements to produce the equivalent of the camera's scan, and how to make light that lasts only as long as human perception.

Braun's cathode-ray tube produces the beam of electrons. Given an input of electric current, it produces a beam whose intensity is analogous to the intensity of the input electric current. The beam is or can be quite thin, like a pencil point. It can be controlled by a pair of deflectors that create a magnetic field through which the beam passes and by varying the intensity of the field, the beam is deflected one way or the other, as in figure 10.6. Two sets of deflectors are used, one for up-down motion, one for left-right motion. Then, we need a timer synchronized with the timing of the camera's scan, something that moves the beam from left to right and at the end of the line moves down one line and back to the left side of the display area, like a typewriter.

Some readers may not remember or may never have used a typewriter. A carriage held the paper and as a key was struck, a hammer with the selected character image at its head moved, struck the page through an inked ribbon, imprinting an image of the character at the end of the hammer head. Then the carriage moved one space to the left to be ready to receive the next character to the right of the previous one. At the end of the line, the typist had to push a carriage return lever which both positioned the page to start again at the left margin and moved it up to start on a new line. In electric typewriters this function could be done automatically. Returning the electron beam from right side to left and down one line serves a similar function.

The earliest experimental televisions worked at about 30 lines per picture. That does not give a very fine reproduction. Imagine

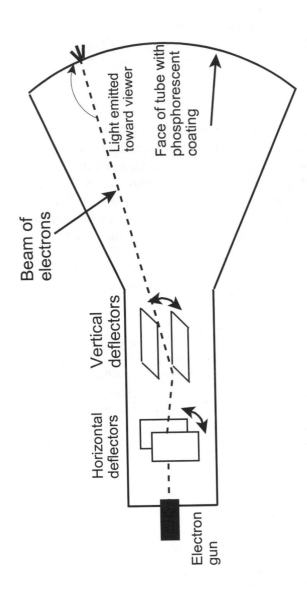

Figure 10.6. *The cathode-ray tube.* Invented by Karl Braun in 1897, this device became and remains the essence of television display. A thin beam of electrons is generated at one end. It passes through two sets of deflector plates, one horizontal, one vertical, and then strikes the back of the face of the tube which is coated with a phosphorescent material. The deflector plates move the beam left to right, down and up, following the pattern of the camera's scanning. The phosphorescent material glows for 1/30 of a second.

a newspaper photograph that used only 30 dots across the page and 30 lines, as shown approximated in figure 10.7. Modern North American televisions work at 525 lines, European TV at 625. The two are not compatible.

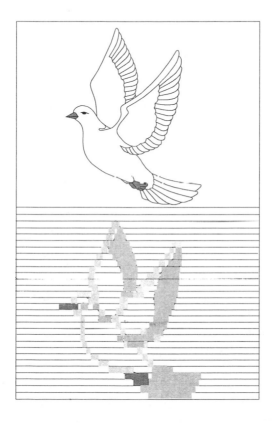

Figure 10.7. *A 30-line image.* The picture below is approximately what the one above would look like if scanned by a 30-line, black-and-white television camera. Stand back far enough from this image and, like the enlarged pixels of figure 10.4, it makes some sense but would never do for modern commercial television.

There are a number of substances that will produce light and glow for a short time when impinged upon by electrons. The phenomenon is called *electroluminescence* or *phosphorescence* and is essentially the opposite of the photoelectric effect. When struck by an electron beam, these substances produce a small amount of light. By selecting just the right chemical material as a coating on the inside of the tube face, the light can be made to persist for 1/30 of a second. In effect, the electron beam paints the picture that the camera saw by quickly lighting up small portions of the face of the CRT, the brightness depending on the brightness of the original scene. Then the picture disappears after 1/30 of a second and the screen is ready for the next image. And *voilà*, we have a working television.

The Developers[8]

In their excellent history of the invention, Fisher and Fisher pose the question of who invented television and answer that nobody knows. They identify four fathers. This was not a team of four; it was four different people, each with his own team, occasionally having some contact with others of the four. They were Paul Nipkow, Campbell Swinton, John Logie Baird, Charles Francis Jenkins, Philo T. Farnsworth, Vladimir K. Zworykin, and David Sarnoff. That makes seven of course, not four. The reason is that Nipkow, Swinton, and Sarnoff were not actual developers of television systems but were of great importance to its development, so we've taken the liberty of adding to the Fishers' four. There were many others but none that made practical, commercially successful machines. The team of Zworykin and Sarnoff brought us to television's golden age. Is it golden, or is TV the "vast wasteland" as it was once characterized by Newton Minow, then chairman of the Federal Communications Commission? We'll discuss this later, but don't look for a final answer. Many have tried to produce the ultimate characterization of television; none has done so. It remains a fascinating and challenging enigma.

In the meantime, let's look at the people who built it. Swinton was mentioned earlier, a man with the right idea but one he did not bring to reality.

Paul Nipkow (Germany, 1860-1940) developed the first camera-like device for television in 1884. What was needed was something that could scan the image and produce first a narrow beam of light, then the corresponding current. He developed the disk with a series of rectangular holes, roughly as was shown in figure 10.5. That this was done in 1884 is amazing, considering that Marconi's early radio work had not yet started. Even if Nipkow had been completely successful, he would not have had radio as a means of transmitting images, and even telephone was still in its early stages.

Nipkow also had an idea for a display unit. It would have been an array of lights, successively turned on or off as the scanning beam moved across the image. This would have been something like C. M.'s idea for a telegraph (chapter 7) that required a separate transmission line and display light for each character that might be transmitted. Nipkow would have had a separate light for each segment of a line, essentially a pixel, that was sent. He never did realize a working TV. Remarkably, though, his disk was still being used in the 1930s, although well before that, both Farnsworth and Zworykin realized that commercial success ultimately depended upon developing Swinton's idea of an all-electronic system that would be more reliable and produce a sharper image.

John Logie Baird (Scotland, 1888-1945) began working mainly with mechanical systems, based on the Nipkow disk and a sort of reverse disk for display purposes. As information from the scan line was transmitted to the receiving disk, the light was projected onto a screen. His system was rather primitive but it did work, up to a point. He sent his first still image in 1923 and gave his first public demonstration of moving images in 1925. In 1929, with the help of the British Post Office of which the British Broadcasting Corporation was then a part, he was allowed to do some test broadcasting through the BBC's radio facilities and thereby became the first to produce true television broadcasting. (Others had sent out images, but his was the first through an established broadcaster.)

Baird was almost always short of funds, so improvements over his initial apparatus came slowly. As competitors, mainly Farnsworth and Zworykin, came on the scene, Baird doggedly stuck to his mechanical system and that eventually did him in, although he did join forces with Farnsworth, if only briefly.

Charles Francis Jenkins (United States, 1867-1934) did some early work on motion-picture projection and his earliest television work was directed at transmission of both still and motion-picture film images rather than live scenes. He, too, produced a working system and did some early experimental broadcasting. His first still image was transmitted in 1922 and first moving images in 1925. He was given a license for scheduled experimental programs in 1928 by the Federal Radio Commission. He went so far as to market plans for converting a radio receiver into a television receiver, using only readily available materials. He, like the others at the time, used the AM radio band for transmitting TV images. He never quite made it to the top. Eventually, his company was bought out by de Forest's company and that, eventually, went to RCA. He and Baird worked essentially in parallel, Jenkins having a slight edge on first transmission, Baird on the first formally sanctioned, scheduled broadcasting. Claims in this history require careful definition of exactly what kind of invention or transmission is being claimed.

Philo T. Farnsworth (United States, 1907-1971) was well deserving of the title "boy genius." He became fascinated with the idea of television as a teenager, read what he could, got some, but little university education. He was obsessed with building an electronic scanner and was convinced that successful television meant all-electronic television, no mechanical parts, hence no Nipkow disk. He built a device that he called the *image dissector*, to serve as the camera. This was a machine that dissected or broke down an image into constituent lines. Like many in this business, he was almost always short of money, but did manage to get funding from the Crocker First National Bank in California. He had no idea how much it would really cost and tended to underestimate, which did not endear him to his sponsors. His first offer from Crocker was for twenty-five thousand dollars. On the other hand, had he known and said what it would ultimately take, he might not have found any sponsors. Probably no one could have foreseen the earnings potential of this new invention at the time, the 1920s.

Farnsworth patented his first television system in 1927, when he was twenty years old, and his image dissector in 1933. All the time he was working there were also others working at AT&T, General Electric, and Westinghouse. AT&T's early interest was in television as an adjunct to the telephone, not as a broadcast medium. When RCA was formed, they became the main competitor. There were visits among these companies, to see what the others had, possibly to buy them out. Farnsworth rejected a buy-out offer from RCA, fearing he would lose control of his work. He eventually signed with Philco Corp., a Philadelphia company that had recently been formed out of the Philadelphia Storage Battery Company and which became a major producer of radios and TV receivers in its day, which is now past. They gave Farnsworth some much needed money but in the long run, it was not a successful partnership.

By 1933, back in England, Baird was put into a competition by BBC. He was to compete against Electric Music Industries (EMI), formed out of the British Marconi company, and therefore now tied in with RCA. Baird had going for him the fact that he was British (this was the BBC, after all) and he had an arrangement with Farnsworth to use some of the latter's electronic equipment. But EMI won out. It was the end of Baird, and Farnsworth never reached the heights he had aimed for, largely because the RCA star was rising so fast.

Vladimir K. Zworykin (born in Russia, 1889, died in U.S.A., 1982) was educated in Russia and did his early TV work there. He studied under Boris Rosing who began assembling a television system in 1907. By 1911 he had produced a "distinct image" of four differently shaded bands. In one sense, this qualifies him as the inventor of television but this primitive image was hardly the basis for broadcast television. Rosing and his star student, Zworykin, got no further at the time.

After the 1917 revolution in Russia, Zworykin came to the United States and soon found work with Westinghouse. He, too, wanted to build an all-electronic television system. During the period when RCA was first formed, Westinghouse was a part owner, as was GE. But, when RCA broke off dealings with these other companies under Justice Department antitrust pressure, most television work of the three ended up with RCA.

Zworykin's first major development was the *kinescope*, a display tube, based on a cathode ray tube. At that time Farnsworth was ahead in terms of image scanning, with his image dissector, but Zworykin developed an *iconoscope* to do that job. He and Farnsworth had a number of patent battles, with Farnsworth getting the nod for priority of invention. But Zworykin had one major advantage over all competition: David Sarnoff.

Sarnoff's early life and the beginning of his career at RCA were described in chapter 9. Just as he foresaw the value of broadcast radio at a time when wireless telegraphy was doing well in the market, he foresaw the potential of television just as broadcast radio was catching on. The timing in both cases was not ideal. His plans for radio were interrupted by the First World War when the energies of Marconi and other companies in the industry were needed for the war. Television went from a laboratory curiosity to a practical, if still technically imperfect, but expensive reality during the Great Depression of the 1930s. As it came closer to fruition, so did the Second World War.

Sarnoff met Zworykin when the latter was working at Westinghouse, a company that had sponsored some early experiments, then given up on television. But Sarnoff arranged to pay for Zworykin's work and later, when the antitrust suit against RCA was settled, Zworykin came over to work for RCA. Even while Zworykin's work was going well, Sarnoff offered to buy out Farnsworth, who refused. When Zworykin first came to RCA, his estimate to bring television to a reality was $100,000, a huge sum in those days. Farnsworth, recall, got an initial $25,000 from his backer, the Crocker Bank. Both overran the estimates. The eventual cost to RCA was actually in the millions, the benefits far higher.

Zworykin was a brilliant man who had a superb team. It should not detract from his accomplishments that he had the moral and financial backing of the most powerful man in the industry. He produced the technology. Sarnoff also realized that to make the new medium successful he needed not only cameras, transmission facilities, and display devices but large numbers of receivers in homes and broadcasting facilities to produce shows.

Broadcasting[9]

We can view the history of television broadcasting as a play in three acts. Act I encompassed the period from the earliest nineteenth-century experiments, notably Nipkow's, to the first successful transmissions in the 1920s. It was exciting for those directly involved, hardly noticed by the world at large. Act II began in 1929 when the BBC began experimental broadcasting to the public. From then until 1939, the public and industry became increasingly aware of this new thing and there was much jockeying for position in what was correctly assumed to become a great industry but one which had not yet shown dramatic growth. Act III began just after World War II when the technical ability to transmit and receive and availability of receivers made the market fairly jump. Act III continued to see improvements, and we are probably still in it. We do not know how this play will end if, indeed, it will. Probably, there will be an Act IV in which television and computer transmission of images merge into a single medium, the computer serving as TV receiver or the receiver as a computer. This capability exists today but is not yet the dominant mode of communicating moving images.

Act I was covered above. In act II we find that Jenkins, Farnsworth, and RCA's new offspring, the National Broadcasting Company (NBC), have begun broadcasting in the United States and Baird has begun his work with the BBC. Later, Germany begins broadcasts. Each country's press or government propagandists claim their own to be first, paying no attention to developments in other countries. Early TV, recall, used around 30 lines per picture, with variations among developers. This was increased by all competitors until around 400 lines were reached.

By the late 1930s there are several companies making and selling TV receivers, some names still familiar today: RCA, Capehart (eventually bought out by Farnsworth), DuMont, Philco, and Zenith. Most of these companies and some others are conducting broadcasts, still not sanctioned by the government for commercial use. RCA announces it will broadcast the opening of the New York World's Fair of 1939, through NBC. They do so, on April 30 of that year, featuring President Franklin Roosevelt's first appearance on the medium. The *New York Times* compares this to the 1921 radio broadcast of Secretary Hoover. In neither case is it the first actual transmission, but both are the most public to date.

A barrier to full commercial exploitation of television is the lack of a transmission standard. Buyers of receivers cannot be sure they could pick up transmissions from all broadcasters. In 1934 the Federal Communications Commission is created and they set up the National Television System Committee (NTSC) to investigate and report on the standards issue. RCA wants to standardize transmission based on its usage of 441 lines. Almost naturally, there is a great hue and cry from the competition since basing a standard on RCA's method would give them a huge advantage in the market. RCA relaxes a bit and the NTSC eventually recommends the 525-line standard we use today. The FCC sets July 1, 1941, as the starting date for commercial broadcasting. At that time receivers are selling in retail stores for prices ranging from $199.50 to $1000. That sounds rather like today's prices, but in 1941 we are just coming out of the depression and average U.S. family income in that year is only $2437.[10]

In the United States commercial broadcasting means advertising. NBC carries one of the first, showing the face of a Bulova watch on screen for a full minute. The cost to Bulova is four dollars for the minute. (The cost for a minute in NBC's broadcast of the 1996 Super Bowl was $2.4 million.[11] Progress.)

In September 1939 World War II begins in Europe. In December 1941 the United States enters the war, just as television is really getting started. Civilian production of radio and television equipment is suspended as the nation gears up for war. End of act II.

Act III begins in 1945. The war has ended, industries are looking for civilian products. FM radio has been developed and television is ready to be mass marketed. People have money.

The sale of black-and-white televisions fairly explodes in the immediate postwar years. Figure 10.8 shows the number of television and cable stations operating in the United States over a number of years. You can see the flatness of the curve during World War II (which was also shortly after commercial broadcasting began), rapid growth just after the war, slowing again during the Korean War (1950-53), and another rapid acceleration just after. Cable shows fairly steady growth. There are more of them than over-the-air broadcasters. Many are quite small.

RCA's National Broadcasting Company (NBC) and its main competitors, the Columbia Broadcasting System (CBS), the American Broadcasting Company (ABC, formerly part of NBC) and now Fox (created by the 20th Century Fox motion picture company) dominate the traditional, over-the-air televison market. CNN and a host of other cable-only channels now compete with the original four.

Color television is first demonstrated by Baird in 1928 but it is not until the late 1940s that it seems ready for full commercial use. The systems had to be robust and the very early ones are not. Also, there is a great controversy over compatibility. CBS, not previously an equipment manufacturer, sets Peter Goldmark (Born in Hungary, then moved to the United States, 1906-1977) to work, and he develops a system they think is ready for the market and to get

FCC approval in 1950 although it is not fully compatible. Compatibility in this sense means that a black-and-white receiver should be able to receive and show color transmissions in black and white. An extra attachment to the receiver is needed for the CBS system to do this.

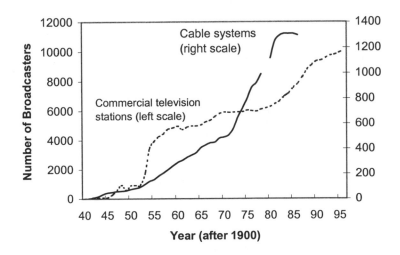

Figure 10.8. *The number of commercial television broadcasting stations and cable systems in the United States, 1941-1996.* Broadcasting grew slowly from 1940, took off after the war, was slowed again by another war in 1950, then grew explosively. Cable has been more steady. There are more cable systems than broadcast stations, some quite small.

Also in 1950 along comes another war, the Korean "police action," and another restriction on some civilian production. RCA does not mind. They are making good money selling black-and-white sets and do not want to abandon this market. Further, they want to wait for a fully compatible color system, which they are to finally develop in 1953. Consumers are protected from the need to scrap their current sets. With that development, CBS abandons its add-on approach and the FCC reverses its initial ruling that had favored the CBS approach and RCA emerges victorious from the color TV war.

Color televisions sell well in the late 1950s, begin to dominate sales in the 1960s, and are the standard by the 1970s. The whole field of television, production of shows, transmission, and manufacturing of receivers, is now a major industry, but it faces new challenges, as have all communications media. These may mark the beginning of a fourth act.

New Developments in Television

Black-and-white television went from around 30 lines per picture in the early, experimental days, to the North American standard of 525 lines when television became fully commercial and government licensed. The sound part of a transmission is sent separately as FM radio. All that required a great increase in the amount of information transmitted, resulting in moving TV transmission from its early use of the AM radio frequency band to the FM band, of much higher frequency. TV stations are each given a band of six megahertz (6 MHz) within which to broadcast, within the region of 54-216 MHz for VHF (broadcast channels 2-13) and 470-800 MHz for UHF (channels 14-69).

Camera Work

Television broadcasting has become quite sophisticated. It is no longer a matter of a single camera pointed at an actor or news reader. There are now many cameras and they are able to function in the studio or outside. Figure 10.9 shows the control room in the

Canadian Broadcasting Corporation studios in Toronto. The director and various technicians monitor a number of possible camera inputs (*feeds*) to decide which shall be shown to the audience at any time. Thus, television is itself a tool of television broadcasting. Figure 10.10 shows a mobile TV vehicle, equipped with a camera (not visible) and an antenna that can send the signal from the camera back to the studio via satellite. This is what gives us the on-the-spot views of ongoing events.

Figure 10.9. *The control room in a modern television studio.* The director and technicians receive feeds from multiple cameras and can choose which to display to viewers at any time. Photo courtesy Canadian Broadcasting Corporation Still Photo Collection, Toronto.

Color

The first major change, once multichannel broadcasting began in the 1940s, was the introduction of color. You can't see color on a black-and-white receiver, so as noted earlier, either the receiver

Figure 10.10. *A mobile television unit.* Contained in a small truck, this unit can go to the scene of some live action, bring out its camera, and transmit the images back to the studio as live action. Photo by C. Meadow.

had to be modified or some form of compatible transmission worked out, which eventually RCA did.

In black and white transmissions, a signal is sent telling how white or dark a specific spot within the image is. In printing with ink, a color image is broken down into three primary colors, using filters to extract the red, blue, and yellow components of the color at any spot. Then a separate printing is made in each of the three colors and sometimes one in black. As the inks mix, the full range of color is produced. To transmit color by television, the original multicolor scene is broken down by filters into three colors, this time red, blue, and green (figure 10.11). No yellow—the manner of mixing is different than in printing but the results are the same. Information about each color is transmitted separately, requiring therefore three times the bandwidth of black-and-white transmissions.

To many viewers, the switch to color was a switch to a different medium. Pictures could be more realistic, advertising more attention grabbing. Until the 1970s I had never seen a hockey game, not even on television, but my then newly adopted city was competing for the Stanley Cup, so I decided to see what it was all about. On our old black-and-white set, I saw images of skaters racing first one way, then the other, seemingly without reason. I could make no sense of it. I couldn't see the puck and sometimes could not tell one team from the other. A neighbor invited me to watch on his color set and I saw a completely different situation. I could clearly differentiate players from the two teams and I could see that little black object that caused all the fuss. News stories of the Vietnam war, in color, were far more vivid than any war coverage had ever been before. Drama could be more dramatic. Animated films could be more fun. What was happening was that TV had become hotter, in the McLuhan sense. Less imagination on the part of the viewer was called for. The machine did more work for us in visualizing the scenes presented.

Cable and Satellite Television

One of the problems of television transmission from the top of a tower to a receiving audience within about 50-80 km is that mountains and tall buildings can block a signal or bounce it back. This makes some receivers unable to pick up some channels and others to get more than one version of an image, one of them a "ghost image," typically offset from the primary one on the screen by as much as several millimeters. A solution, beginning in 1948,

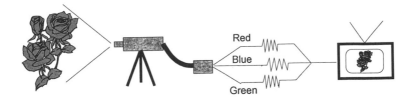

Figure 10.11. *Separation of the color elements of a television image.* A filter separates the image into three colors, red, blue, and green. These are separately transmitted, then recombined at the receiving end.

was to set up one highly elevated receiving antenna for a community, with cables that ran from the antenna to each subscriber's home or office. Such an arrangement was called *community antenna television* (CATV).

All sorts of predictions were made for how this new medium would change our lives. It gave the capacity for more channels than we could get over-the-air and locally originated broadcasts would not cost as much as if each broadcaster had to provide complete transmission facilities. It could provide education and community information and make for more democracy in our communities. There *is* some educational and local programming, but in truth the net effect has largely been to provide higher quality reception and that has led to the use of what is now called simply *cable television* in many areas, not merely those hampered by interference with over-the-air signals. The cable brings us more channels. We do get more variety and that has provided for some of the predicted social benefits, but no revolutionary change.

More recently transmission via satellite has begun to challenge the cable providers. In effect, the transmitting antenna is moved up to a satellite, 37,000 kilometers above the earth, and it broadcasts down to a wide audience below. These systems offer to provide even more channels than cable, although not all areas have yet seen any dramatic increase. Their main advantage is that they do not require terrestrial cabling. They do require a receiving antenna in each building, but these have shrunk in size from around five meters in diameter to as small as half a meter.

Claims and counterclaims are being exchanged by the respective providers. Since most large cities are already cabled, that industry has a great advantage. If the 500-channel world would actually come about, it could well affect the balance of power. Some cable and satellite operators are now offering Internet access, as well as television, giving faster transmission speeds than have traditionally been possible through telephone connections. This adds considerable value to the transmission system.

Yet another aspect of the competition between cable and satellite as television delivery systems is fiber optic cable. This is a relatively new technology, to be discussed in chapter 12. These new cables offer a very high bandwidth. At this time, these cables do not generally reach individual households, as do twisted-pair wires and televison cables. But it has become a common sight to see streets being torn up to allow for running fiber optic cables under them and in a few years we might find them reaching our individual homes and offices. The effect will be to be able to bring in far more television channels, shopping services, voice telephone over the Internet, and who knows what else. It will take a great deal of message traffic to fully utilize these cables. At this time, even the outcome of the commercial satellite versus cable war is in doubt and getting fiber optics into the home is still a dream.

Video Recording[12]

Video recording (VR) means recording a portion of a television transmission, with full motion and sound. The transmission can be of a broadcast program, a studio production, or a home video. Like writing or drawing, VR is not a transmission system in the usual sense. It is a means of recording for later use, or communicating over time. Indeed, some of its earliest developers saw it as primarily a way to defer watching a broadcast, or *time-shifting*, rather than to record motion pictures and thereby compete with theater showings.

In the story of this invention we cannot identify an individual as the principal developer. We attribute its development to corporations and sometimes it is even difficult to be sure which corporation gets the principal credit. The first successful recorders were built in 1956 by Ampex Corporation, an American company, and were intended primarily for use by the television production industry. Earlier attempts had been made by RCA and Bing Crosby Enterprises, owned by the singer Bing Crosby, but they had proved unsuccessful. The recording medium was magnetic tape. Then SONY Corporation in Japan became interested in adapting the concept for consumer use. After several trials and working in cooperation with several other Japanese companies the *betamax* recording system was developed in 1975. Betamax was small enough to be a home product although expensive, initially priced at over one thousand dollars. It used a compact tape cassette, and it could record for an hour.

Some of SONY's partners, led by JVC (originally Japan Victor Corporation, now just known by the initials) came to feel that consumers wanted more recording time, even if it required a larger cassette. In Japan, it is common for companies to share both development costs and patents. Each side in this dispute had its supporters but JVC won over most of the partners, creating a new system called *video home recording* (VHS) and by 1978 this became a de facto standard, that is unofficial but generally accepted. It remains so today although both systems gradually improved their maximum recording times, essentially nullifying the original VHS advantage, but VHS by then dominated the home recording market, and prerecorded programs or movies in that format were becoming popular.

There remains some controversy about which system was better and how VHS eventually won. We cannot resolve the question but it seems that both standards had their advantages, neither clearly better on all counts. Once VHS became popular it was impossible to unseat it as the market leader.

Around the same time as video-tape recording was being developed, some companies were experimenting with recording on disks, called *video disk recording*, using lasers. Although disk-based systems were made to work, the problem was that recording had to be done in a factory or at any rate with a special machine. The average television viewer could not make video disk recordings at home. If they were prerecorded they offered the advantage of greater reliability and longer life. In 1981 RCA brought out *selectavision*, a disk reader using a ten-inch disk, but withdrew it from the market in 1984. VHS had won again.

The idea of disk recording would not die. In 1985 the read-only compact disk (CD-ROM) for computer use came out, and this evolved into a recording form that could be used for recording at a density high enough to hold a commercial motion picture, typically about two hours. The basis was the DVD described in chapter 3, now produced in a variety of formats depending on intended use, such as computer memory, audio recording or video recording.

The video DVD was originally developed by a partnership of SONY and the Dutch firm Philips Electronics N.V. It quickly became an industry standard. There is now a larger partnership, the DVD Forum, that takes care of standards and patent licensing.

Today, VHS tape is still widely used with home video recorders/players. Since DVDs cannot be recorded on a home machine a separate reader is needed for them, but an increasing number of motion pictures are coming out in this longer lasting form. Since computers can now record on a form of compact disk, it seems clear that a home system able to both play back from and record on DVDs will be upon us soon, and VHS will disappear. We can now record on CDs with a computer, suggesting that recording of TV programs at home may come soon.

Digital Television[13]

Digital television (DTV) is a transmission system, not a new kind of television. By converting camera images to digital representation and transmitting them digitally, broadcasters can reduce the effect of various forms of noise which degrade the received images. At the viewer's set, the signals are restored to analog and that is how the receiver processes them, regardless of how transmitted. There is a parallel to music transmission by radio. The transmission may be digital, but what the listener hears is the result of converting digital to analog sound of high quality.

In the year 2000 DTV was largely sent by satellite. Their customers need a converter anyway, to receive satellite signals, so there is no extra cost if the signals are digital. Some cable operators have begun to change transmission modes or to offer the user a choice of analog or digital modes, but users are not pushed into switching by resolution-destroying noise that interferes with over-the-air broadcasting. Typically, they are charged extra for a digital-to-analog converter.

High Definition Television[14]

What the television world is now waiting for, some eagerly and some with dread, is a new transmission system called *high definition television*, or HDTV. In one sense, it doesn't sound like much. There will be an increase from 525 to 1125 lines, cutting in half the thickness of a scan line on a TV screen thereby making the image much sharper. The relative dimensions of the screen will change, making the TV image look like a smaller version of a wide-screen motion picture image, as in figure 10.12. In today's receivers, to give the impression of a wide screen, the top and bottom of the viewing screen are blacked out, giving an image that seems wider but shorter.

Those who have seen HDTV are generally effusive in its praise. I am among them. When I first saw it demonstrated sports, football and swimming, were most dramatically different. We watchers had a view of a football game not available in the stadium. We had sharp, close-up images, seemingly closer than is possible at a stadium and sharper than on conventional TV. We also saw a sharp close-up of a swimmer doing the butterfly stroke, pushing a wave in front of his face, yet we could see the face clearly through the water. The drama we saw was a quite conventional shoot-'em-up. There is just so much you can do to enhance the image of a building exploding. It is a new medium. It will be up to screenwriters, set designers, and directors to learn to make maximum use of the new capabilities and not just show explosions with higher resolution. Will there be some compatible transmission system that allows us to pick up 1125-line transmissions on a 525-line receiver? One proposal is that transmissions be sent simultaneously over both newly assigned very-high-frequency channels as well as in the old format over the old channels for some years to come, until it can safely be assumed that nearly everyone has the new equipment. Will the images we see in bars or other peoples' houses (that's how color was introduced, most people went out to see it at first) convince us to scrap our present systems? Will the FCC license an incompatible system? How would a yet hotter form of TV affect programming? No one knows yet. If we could have it for free, we'd probably all like it, but that won't happen. As a new medium, how will it be used? I doubt that anyone has ever successfully predicted the effect of a new communications medium. If no one else gains from this one, the advertising companies hawking the new equipment surely will.

Another aspect of the new technological capability to compact the transmitted image would be to use new channels, not to give greater definition, but to send more than one conventional TV program in the assigned frequency range. This would cut costs for broadcasters and offer up to twice as many channels for viewers to choose among. Tune in again in five or ten years and see what happens.

Figure 10.12. *The new shape of television?* A comparison of the aspect ratios of present-day television and the proposed new high-definition television, the latter approximating wide screen motion picture displays. The image above shows a scene as it would look on HDTV (ratio of width to height about 1.8). The lower image, assuming the same screen height, shows what we would see on a conventional television, aspect ratio of 1.3. Photo by C. Meadow

Interaction

The concept of interactive television has been with us since community antennas were introduced. The idea is to make TV into two-way communication.[15] This can't be done with conventional, over-the-air broadcasting because there is no way to get the viewer's signals back to the broadcaster. But with cable, there is ample bandwidth, even though some form of transmitter would be necessary at the viewing end. The original ideas were somewhat less than enchanting and they went nowhere. Proposals included letting viewers suggest a direction for a discussion panel to take or ask for more information about an item on the news. There were speculations, to my knowledge never carried out, about having the audience determine the ending of a drama or make coaching decisions in a ball game.

Today, talk of interactive TV is being renewed. Now that so many people have access to the Internet it is possible to use either the computer in the home or an inexpensive attachment to the TV cable to talk back to a broadcaster. Asking for more information remains one of the attractions. The replies could be delivered on some channel other than the one carrying the TV broadcast, so only the requester would see it. Other possibilities include electronic shopping or *e-commerce* where a viewer could ask for more information about products and receive it on the computer. This capability might make this new form of commerce more attractive than current television shopping. Live conversations with sales or customer service people can be carried out as well. Yet another possibility is for the viewer to order a shirt just like the one the hero is wearing in the movie on view. The possibilities for the commercial aspects of drama to dominate the script are depressing. We'll have more to say about this later.

A device called a *set top box*[16] will be helping to achieve interaction. We all had a set top box in the early days of cable. It was necessary to convert the cable signals for display on our receivers and our remote unit communicated with the box, not the TV receiver, itself. The new box may perform a variety of functions. One is simply to convert incoming digital signals to analog. Another is to carry on the interaction between viewer and TV program. If, for example, the viewer wanted to order a shirt such as the hero is wearing, a signal goes to the box which would relay it to the seller. What about the program while that transaction is going on? One form of box would begin recording the program as soon as the viewer initiates interaction and, when the sale has been consummated, the recorder begins playing back the program. Useful if the interruption is for dinner or a restroom break; maybe not so appealing if the program is showing live news or football.

Computers *versus* Television

The Internet is now capable of delivering motion pictures and although the average computer screen is smaller than the average television screen, a larger monitor can always be attached. This need not affect programming and production (in the theatrical sense) but it could have a dramatic effect on the transmission industry and on the manufacture and sale of television receivers. Some receivers are now being built with the capability to communicate with the Internet, essentially letting the TV encroach on the computer business. This is another area in which predicting the winner is impossible. Actually, we may see some kind of hybrid device that is both computer and TV or see even personal computers become little more than communications devices, with all computing done and data stored at a site belonging to an Internet service provider. My head hurts.

Computers *with* Television[17]

A new technological development is the ease with which virtual images can be combined with actual images by computer manipulation. This is called L-VIS, for *live video insertion system*. The camera takes in a scene but a computer modifies it in such a way that the viewer does not know which images are real and which were added by the computer. This was done quite skillfully in the movie *Forrest Gump*, where the hero was shown in a realistic-looking scene shaking hands with President Kennedy, who was already dead when the movie was made. Images of the actor were inserted into real, old newsreel pictures. But this was fiction. Two simple examples from televised sports seem equally harmless.

There was a brief fling with inserting a processor chip in a hockey puck, causing it to seem to leave a purple wake behind it when seen on TV. This was derided by hardcore hockey fans. In football, a yellow line was added as if it were a yard line on the field. It showed the line the offensive team had to reach in order to make a first down. Again, this line is not real, not actually on the field. It is intriguing that as a real player crosses this virtual line, his image obscures it, just as if it were really painted on the field, so it is not as simple as overlaying one image on another. The yellow line overlays the green grass but the players overlay the yellow line.

Figure 10.13. shows a portion of a televised image from the 2001 Super Bowl football game. The imposed yellow line can be seen as gray at **a** in the figure. The regular white line is just to the left of the letter **b**. At **c** is an inserted advertisement, obscured as players pass over it.

At the New Year's Eve celebration of the new millennium (December 31, 1999) in New York's Times Square, the CBS camera had an NBC advertising sign in its view, a real advertising sign. CBS electronically inserted their own logo over what was supposed to be a real picture. The problem is, what if they or some network starts to do this with serious news? Can the viewing public accept the integrity of televised images? This is potentially a powerful technique for presenting televised or cinematic fiction. It must be used responsibly.

Figure 10.13. *Virtual television images*. This image was made from a televised football game. It shows two kinds of images added to the scene viewed through a camera at the stadium. In the upper left is a CBS logo and the score of the game. This is simply superimposed on whatever the camera sees. To the right of center are two lines, marked **a** and **b**. Line **b** is white and is actually marked on the field of play, hence visible to the live audience. Line **a** is added in the studio. It is yellow on the screen and not visible to people at the stadium. Note that when a player crosses this invisible line, he obscures it, unlike what happens with the score inset. The advertisement at **c**, lower center, is also added, not on the field, and obscured when players or observers seem to cross it. These techniques are capable of making considerable changes in what appears to be a live image. Photo courtesy Princeton Video Image Inc.

Impact

In its short life, television has been a constant subject of debate as to its impact on society. Does it rot our children's minds? Is it killing interest and skill in reading and writing? Has it killed the art of conversation? Or has it been a way to inform society about what is happening in the world, beyond what any other communications medium has ever before been able to do? If this last is true it would probably be because of television's ability to hold an audience's attention and not to demand much in the way of imagination by the viewer. Granted, not a great deal of deep material is presented in the average news program, but something is. We were never as close to our political candidates before TV, but the new closeness has not turned out to be much of a benefit as we find political candidates evermore concentrating on their TV images rather than the substance of their political or economic plans. We witness many more events than we ever could have before: a royal wedding; a presidential funeral; the bombing of a city televised from within the city; the scene of a mass killing; or the murder trial of a famous person.

We also see fiction that appears to many of us as tasteless. We see both fiction and news presented as a series of short, high-intensity bites. On the other side, programs such as *Sesame Street*, *Mr. Rogers' Neighborhood*, and *Arthur* have been generally highly regarded fare for the very young.

The strongest criticism of television seems to come from the most literate observers, those most devoted to the printed word. Live theater, well regarded in ancient Greece when it was the only game in town, is no longer so highly regarded in Europe and America where the hotter media of motion pictures and television have won over audiences. Thomas Hardy's 1872 novel, *Under the Greenwood Tree*, told about the human interaction problems that arose as a result of bringing an organ into a British church previously accustomed only to an unaccompanied choir. Here was a machine being used to make the music that only people had made before. And, of course, there was only one organist needed but many choristers. Introducing new communications media can be a highly disruptive process, even if it improves the quality of the content.

We must remember, in judging this medium, that the quality of programs presented is not inherent in the medium of transmission. There is nothing fundamentally wrong with television as a technology. If there is something wrong, it tends to be with the programming presented, i.e., with the producing medium. And since most of television in North America is commercial, programming is based on what sells and what sells to most people does not always fit the tastes of the most literate segment of the population. We should also remember that the thinness of television news is based on achieving wide audience appeal. Is the general population better informed on important issues of the day than they were before television? Before TV, did most people get their state-of-the-world information from the *New York Times*, *Foreign Affairs*, and the *Wall Street Journal*? Almost certainly not, and almost certainly they were not as well informed as they are today, even if they could still be far more so. Still, television reduces our need for analysis and imagination. It os hard to see how this will be an advantage to people raised on it.

When television became a practical reality in the late 1940s, in a sense the world was not ready for it. In a sense the world has not been ready for any of the major new media but here was one that could do so much in art, entertainment, and education, and offer so much profit for the entrepreneurs that maybe we never took the time to learn how best to use it. Technologically, it has been a superb accomplishment.

The questions we have raised are hard to answer, made all the harder because we really do not have a comparable civilization without television to compare with. There were those who felt the printed book was a cultural horror. In his discussion of Homer's *Odyssey,* Bernard Knox reminds us that "Early printers tried to make their books look like handwritten manuscripts because in scholarly circles printed books were regarded as vulgar and inferior products."[18] We still see something like this today: modern American telephones shaped in the style of century-old European ones, replicas of 1930s radios, twenty-first century automobiles styled like those of the 1930s. Old classics of any objects seem to retain a sense of value beyond that of the new ones. Perhaps that is what makes for disdain in some circles for the newly rich, compared to families with "old" money. Books with leather bindings—common a century ago—are treated with near reverence today, even if unread. The old classics maintain an air of respectability long after new ideas, media, or even money are introduced.

Notes

1. Much of this chapter is based on material found in Fisher & Fisher, *Tube*. This is the best single source for further reading. See also Smith, *Television*.
2. Vanderhaeghe, *The Englishman's Boy*, 107.
3. Fisher & Fisher, *Tube*: Swinton's original thoughts, 38, 60, and change of heart, 79.
4. Quote from *Guardian* in Fisher & Fisher, *Tube*, 248.
5. "Facsimile," *Encyclopaedia Britannica;* Carpenter, *Inventors: Profiles*.
6. "Television," *Encyclopaedia Britannica;* Fisher & Fisher, *Tube*.
7. In my early days in computing, it was common to convert the flow of electronic signals, from one part of a computer to another, into sound. Listeners could not tell exactly what was happening, say that we're now adding 2 to the previous total, but we could recognize a sound pattern and often detect that something was wrong.
8. Fisher & Fisher, *Tube*; Smith, *Television*.
9. Inglis, *Behind the Tube*; Rather, *Deadlines and Datelines*; "Broadcasting," *Encyclopaedia Britannica*.
10. *Historical Statistics*, 186.
11. Fisher & Fisher, *Tube*, 296.
12. Lardner, *Fast Forward;* "SONY," *Jones Telecommunications & Multimedia Encyclopedia*.

13. Fisher & Fisher, *Tube,* 340-351; Silbergeld & Pescatore, *Guide*; "Digital Television," *Jones Telecommunications & Multimedia Encyclopedia.*
14. Fisher & Fisher, *Tube, 340-351*; "High-Definition Television."
15. For perhaps the first try at interaction, see Lewis, *The New New Thing*, 72.
16. O'Driscoll, *Essential.*
17. Goddard, "The New Fad."
18. Knox, "Introduction," 4.

11

Communication Satellites[1]

Introduction

A communication satellite is not a new medium. In a sense, like television, it is a new way of using an existing one. It is a space vehicle carrying an antenna and a relay station for microwave communication, many kilometers above the earth's surface. The vehicle is an artificial satellite, often called a *bird*, or a mirror in the sky in orbit above the earth. We can now transmit from any place on earth that has a transmitting antenna to anyplace with a receiving antenna without the need for connecting wires between sender and receiver. The earth keeps shrinking.

The Nature of the Bird[2]

Communication satellites can be placed in *low earth orbit* (LEO) about 160-480 km above earth, *medium earth orbit* (MEO) 960-19,000 km up, or *geosynchronous* or *geostationary earth orbit*, (GEO) 37,000 km or 22,300 miles above the earth. Satellites in GEO can "see" over and be seen from about one-third of the earth's surface. Therefore, the line of sight restriction on high-

frequency electromagnetic wave transmission is not much of a
barrier to long-distance transmission. Three satellites floating (an
inaccurate term) over the equator, 120 degrees apart, can cover the
earth (figure 11.1).

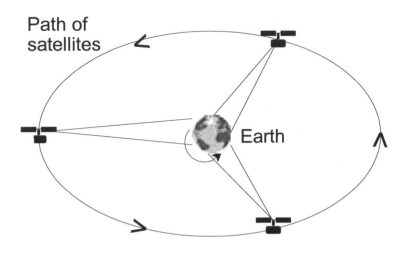

Figure 11.1. *Coverage of the globe by communication satellites.* As few
as three satellites in a high, geosynchronous orbit, can cover just about all
the earth's surface.

The geosynchronous orbit means that the bird, while actually
moving, as viewed from the earth which is also moving, seems to
be standing still. The earth rotates once every 24 hours and the
satellite completes one orbit in the same amount of time, as shown
in figure 11.2. It is like driving in a car at 100 km/hr and noticing
that the car in the next lane, moving at the same speed, seems to be
standing still. For wireless telephone, many satellites in a LEO
orbit appeared more satisfactory than fewer in the higher GEO.
    The earliest nonmilitary vehicles were simply reflectors. That
meant that signals from the earth would have to reach not only the
bird but have enough power that the portion of the signal reflected

would be able to be picked up back on the ground. Today, satellites contain repeaters that regenerate a received signal, much reducing the amount of power needed in a ground transmitter. The need, then, is for a ground station that sends signals up, the bird itself, and another ground station to receive signals, as shown in figure 11.3.

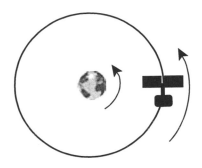

Figure 11.2. *A satellite in geosynchronous orbit.* If positioned at the correct altitude and with correct speed, then as the earth rotates on its axis the satellite moves through its orbit and appears to observers on earth as if standing still. Therefore it is always accessible to transmitting and receiving antennas in the area it covers.

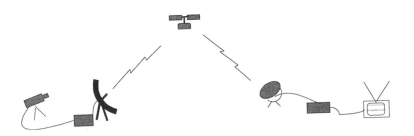

Figure 11.3. *Earth station to earth station linkage.* Here we see the typical steps of a signal going from its earthly origin, shown as a television camera (at left) to a transmitter, to the satellite, back down to a receiving antenna on earth and into a television receiver.

A slight delay is introduced when using a satellite. A signal must go up to the full 37,000 km then go down 37,000 km more. Given that amount of travel, the distance on the ground from the sender to the receiver, say the approximately 1500 km from New York to Chicago, is of little account. But it takes about one-eighth second, to get there and the same to get back down, for a total of nearly a quarter second. That quarter-second total may seem trivial but it can be noticed in conversation. If a continuous television program were being sent, the viewer would see each frame a quarter second after it was sent, the whole program would be uniformly delayed, and the net effect would be hardly noticeable. But John Pierce of Bell Laboratories, one of the pioneers in this field, pointed out that an imposed delay in a telephone conversation of as little as one-tenth of a second seems "unpleasant and upsetting."[3] If telephone were multimedia and persons engaged in a conversation could signal the need for a slight delay by gesture, this might be more tolerable, but the imposed delay can come at any point in a conversation and is unanticipated by the listener.

Communication satellites are used today for radio, television, and telephone which includes much of communication between computers. The radio applications include sending of navigational and meteorological data. The birds reduce the need for terrestrial infrastructures based on wire and cable. They are enabling less developed countries to get the benefits of modern telecommunications without the need for wiring up their nations, and they enable highly technologically developed countries to proliferate the use of wireless telephone. A modern communications satellite is shown in figure 11.4 and an antenna used to send up or receive signals is shown in figure 11.5.

History[4]

The two people who get the most credit for originating the idea were Arthur C. Clarke and John R. Pierce. Clarke (born in Eng-

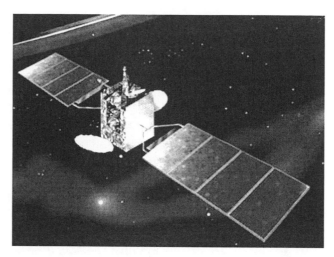

Figure 11.4. *A modern communications satellite.* This is the INTELSAT 805, launched into a geostationary orbit in 1998. Photo courtesy INTELSAT.

Figure 11.5. *A modern antenna for communication with a satellite.* Such antennas range from one-half to thirty meters in diameter, depending on intended use. This is one of the larger ones used for voice, data, or video communication. Photo courtesy INTELSAT.

land 1917, has lived his later life in Sri Lanka) first proposed communication satellites in 1945. While he was best known as a science-fiction writer (most notably *2001: A Space Odyssey*), he was also an electrical engineer. He was in charge of introducing the radar application, Ground Controlled Approach, to the Royal Air Force in 1941.[5] John R. Pierce (United States, born 1910), while director of electronics research at AT&T's Bell Laboratories, independently proposed communication satellites in 1955 without having been aware of Clarke's earlier proposal. The two became personal friends so, in this case there were no patent squabbles.

The germ of the idea and the basic technology predate the involvement of both these men. Pierce credits a science fiction book published in 1911 with the first suggestion and a story in 1942, as well as his own sci-fi story published under a pseudonym in 1952, all preceding his own first scientific paper on the subject.[6]

As early as 1946 astronomers had detected radar signals that went from earth and were bounced off the moon. It was simply that—send out a signal, then recognize its reflection coming back, as with radar. But these were message-bearing signals. This meant that it was possible to use an extraterrestrial object to assist communication between points on earth. The U.S. Navy used the moon in 1959 as a communications satellite linking Washington, D.C., and Hawaii. In that same year Bell Laboratories used it to test facilities later to be involved with its first human-made communications satellite.

The world's first artificial satellite of any kind was the Russian *Sputnik*, launched October 4, 1957. It communicated only by sending a repeated beeping signal back to earth but it worked, it was the first of its kind, and it shocked a western world used to its own technological superiority.

The first artificial communication satellite was built for and operated by the U.S. Air Force. Launched in December 1958, it served to record messages as it passed over a transmitting station and played them back as it passed a receiving station. It stayed in orbit for only twelve days but, again, demonstrated what was possible.

The first civilian communication satellite was called *Echo*. It was a joint project of Bell Laboratories, the Jet Propulsion Laboratory, and the National Aeronautics and Space Administration (NASA). Launched in 1960, it was a balloon, about 30 meters in diameter and coated with a metallic film to reflect microwave signals just as the moon and the airplanes had. The problem with it was that all the power had to come from the terrestrial transmitting station. Even though the beam going up was narrow, there was loss due to spreading and even more when the reflected signals came back. In fact, only one millionth of one millionth of one millionth (10^{-18}) of the original power made it back to a receiving station. But it, too, worked. Further, it was visible with the naked eye, looking like a moving star, and that brought a lot of popular attention to the technology, never a bad thing when funds are short. *Echo* was not geosynchronous and that also limited its usefulness.

The next type, actually a series of satellites, was *Telstar*, a Bell Laboratories undertaking,. The first one was launched in 1962. The *Telstars* had repeaters but were still not in a geosynchronous orbit. That meant they were only visible at any point on earth for a limited time. That, in turn, meant you could not broadcast a baseball game or movie without interruption, but you could send telegram-like messages or hold short voice conversations.

The *Syncom* series came along in 1963. They were in geosynchronous orbit. This system was so successful that commercial exploitation was deemed feasible by the U.S. government which set up the Communication Satellite Corporation (COMSAT) to oversee building and operating of satellites for the United States. Shares were sold to corporate investors, largely AT&T, RCA, International Telephone & Telegraph (IT&T), and General Telephone and Electronics (GTE).

The first live transatlantic television signal was sent via *Telstar* in 1962. The Tokyo Olympics of 1964 were broadcast on TV in North America. How live was coverage of the games when the satellite could not both receive from the ground at Tokyo and transmit to North America simultaneously? By holding the image until the bird passed over North America, or relaying it through ground stations, the images could find their way to American audiences quickly. The concept of "live" in television does not always mean "live" as you're viewing it." The images might be hours old. Today, of course, we routinely see TV news broadcasts originating almost anywhere in the world.

In 1964 INTELSAT was created as a cooperative company owned eventually by 140 countries. It plays, internationally, a similar role to that of AT&T's Long Lines Division before the breakup of that company. It provides carrier services to other telecommunications services that, in turn, connect to consumers. INTELSAT launched its first satellite, *Early Bird*, in 1965, provided worldwide transmissions for the 1969 moon landing, and in 1978 enabled over a billion people in forty-two countries to see the World Cup soccer matches. Today, INTELSAT covers the globe, offering transmission services for voice, data, and television.[7]

Technicalities

A satellite has to be pushed into orbit by a rocket, now a routine occurrence, but not so in the 1950s when just getting the rocket successfully launched was a problem. If in a low orbit, the bird would seem to move relative to the earth and would not see nearly one-third of its surface. One in a higher, geosynchronous orbit could see that far but needed more rocket power to boost it that high.

Satellite communications are carried out at very high frequency, which allows for wide bandwidth and avoids some forms of interference from atmospheric phenomena.

The communication capacity varies considerably, but the *Intelsat VI* satellite (1989) could transmit as many as 35,000 telephone channels (i.e., simultaneous calls), which compares quite favorably with the 4200 available through the transatlantic undersea telephone cable in 1983. Land based cables, grouped into clusters of ten cables, can handle about the same number as satellites. Fiber optic cables have made this form of cable competitive with satellites. Is it better or cheaper for any given user to use a satellite or a fiber cable? Fortunately, consumers deal through a telephone service provider and need not be directly concerned with what device the service provider uses to carry the message.[8] Remembering how much of the earth-generated power was lost in using Echo, the question of a power source has always been of importance. We are now able to put solar energy collectors in even a small satellite and these provide all the power necessary to receive and retransmit signals. Relaying the messages means the signal power going up can be considerably less than was required of Echo. Without this, the telephone systems using satellites would not be possible.

Let's consider three applications of these vehicles: wireless telephone, position finding for navigation, and television transmission.

Wireless Telephone

The Motorola Company, a pioneer in cellular telephone, created the Iridium system in 1998. This consisted of sixty-six vehicles in low earth orbit which were able to communicate directly with small portable telephones. The network of satellites covered the world and enabled any registered user to reach any conventional telephone or other registered user, anywhere in the world. Users would register with a local service provider, not directly with Iridium.[9]

To accomplish what Iridium set out to do, the space vehicles would have been able to communicate with each other if originator and destination were not in communication with the same bird.

Satellites used for telephone service have a system for intercommunicating somewhat like that used for ground-based cellular telephones, which we skipped over lightly in chapter 8. When a cellular telephone user originates a call, it goes to the nearest cell which notes which phone is calling and what the destination is. Then the system has to find the destination. If it is a fixed telephone, a connection can readily be made to the regular telephone network. If, however, the destination is also a cellular phone, the system has to find where that destination phone is. Cell phones are polled every now and then, to determine their location, in case a user is "roaming." The polling is done without the user being actively involved. The potential recipient's location is noted by the system and when a call comes for that phone, the linkage can be made from caller to nearest satellite, to a ground station that has location information, then to the satellite closest to the destination.

It is easy to see that there is a great deal of activity involved in polling and relaying messages and keeping track of where all the subscribers are. This would be impossible if not done automatically by computer.

Unfortunately, the imaginative Iridium system was not financially successful. The service was halted in 2000. The vehicles were then sold to another company for much more limited use. It appears that more localized wireless systems took too many of Iridium's potential customers. Most such services need terrestrial cells or connection points but could connect into the worldwide telephone network if needed. The main beneficiaries of Iridium were miners, prospectors, and others who work in remote areas and need connections to home offices or other remotely located colleagues.

Navigation[10]

Today, not just navigators of ships at sea or pilots of aircraft, but almost anyone can determine his or her current location with the aid of a relatively small bit of hardware and another system of intercommunicating satellites called the Global Positioning System (GPS). This also uses many satellites because a user has to be able to receive information from at least three of them to a get a good navigational fix on his or her location. There are twenty-four vehicles going around the world. By combining information from these vehicles, latitude, longitude, altitude, and current time can be determined.

What is "good" location data? GPS was created by the U.S. Department of Defense, originally for its own use. A using station can determine its position within 22 meters horizontally and 27.7 meters vertically and time within 100 nanoseconds (billionths of a second). For ocean navigation such figures are fully adequate. They are even adequate for automobile navigation, finding your location on an ill-lit country road at night.

Satellite or Direct Television[11]

Television is typically handled by satellites in geosynchronous orbit. In one mode TV "messages" are not sent to individuals but to a broad audience spread over a wide geographic area. For the United States and Canada, a single vehicle could service the entire territory. Or, signals might be sent only to a local cable operator for terrestrial redistribution by that medium. For images sent from one continent to another, relays might be needed.

Although it is possible to relay TV signals from one bird to another, in practice it is more likely that the signal will go up to one vehicle, down to an earth station and then, if necessary, up to another satellite. It's a matter of what gives the most reliable and economical transmission.

We have become used to seeing programs that originate on other continents, mostly news and sports, but sometimes entertainment programs as well. When doing this we encounter the standards problem. Remember that Europe, Asia, and America use different standards for, among other things, the number of scan lines that constitute an image. To communicate from one standard to another it is necessary to convert the form of the image, roughly by deleting lines coming from Europe and adding, interpolating, or duplicating some lines if the transmission originated in the Americas. The other alternative is for the western hemispheric broadcasters to send their own camera crews overseas, using their own equipment from beginning to end. The conversion is the responsibility of the television companies. The satellite operators transmit whatever signals they are given. Making sure they're the right signals is the responsibility of the originator.

Another aspect of satellite-delivered television is its competition with terrestrial cables, mentioned in chapter 8. Starting from scratch, satellite TV seems more economical—no extensive cables to install, the birds are already there, and the viewer's antenna can now be small enough to sit on a windowsill. But, there is much cabling in place, especially in larger cities where the most viewers are concentrated. In less populated areas satellite seems to have a clear edge.

Impact

The world's telecommunication facilities began to grow explosively in the middle of the twentieth century. Much of this was due to tremendous growth in computer-to-computer communication and much to increased use of long-distance facilities, by then no longer reserved for emergencies. Communications satellites offered and delivered the only means of reaching nearly anywhere on earth from nearly any other place on earth. It was quickly realized that nations without a well-developed communications infrastructure, basically meaning telephone wires, could join this communications frenzy. Today, the glamor application seems to be in wireless telephone and its related paging service, enabling the busy person to be ever in touch with family or colleagues.

Certainly, the satellites are contributing to a sense of globalizing the world. We can see what is happening anywhere, as it happens. The social consequences are many but not all benevolent because it means bringing a picture of the rich world to the poor world and that has been known to create great resentment among the have-not people. It has also brought images of the reality of war and famine to people who have never experienced them and while that is not necessarily good, it does make for a greater appreciation of reality.

In my personal experience, the view of the United States that is seen overseas on television is highly distorted. The selection of programs for export (or perhaps it's the selection by receiving countries for import) tends to show a people obsessed with violence or a people who all seem to have everything, in the material sense. (This from having watched American movies on local TV in Colombia and Jamaica. What viewers there could see was not the best of what the U.S.A. produces.) Neither the view of Americans as both rich and violent, nor of a people seeing only moral and enlightening programs is totally correct. How correct is our view of other peoples? How successful have our telecommunications facilities been at showing any people what other peoples are really like?

The ability to keep in touch by cellular or wireless telephone is very comforting when driving alone, away from the city and pay telephones, or when someone is sick, or an important deal is pending. What will be the ultimate effect on society when we can never really get away? Difficult to impossible to tell.

Satellites have been one of our most spectacular accomplishments in telecommunications technology. Less popularly spectacular has been the growth of fiber optic cables, now criss-crossing the oceans. Together, we have an industry that does not just grow, but explodes. There is some danger that, as happened to Iridium, change will come too fast for those investing in expensive new technology ever to realize a return on their investment.

Notes

1. Pierce, *Beginnings*; Clarke, *How the World.*
2. Pierce, *Beginnings*; Clarke, *How the World.*
3. Pierce, *Beginnings*, 30.
4. Pierce, *Beginnings*; Clarke, *How the World.*

5. Clarke, "You're on" in *How the World*, 154-161.
6. The book was Gernsback, *Ralph124C41+*; The early story was Smith, "QRM Interplanetary"and Pierce's own story was "Don't write; telegraph."
7. www.Intelsat.com
8. "Telecommunications Systems," *Encyclopaedia Britannica*.
9. www.Iridium.ca/faqs-frame.htm [Unfortunately, no longer available.]
10. Dana, Peter H. *Global Positioning System Overview*; *Understanding GPS*.
11. Hudson, *Communication Satellites*.

12

The Internet and the
Information Highway[1]

Introduction

There has been a lot of confusion about what the Internet, the
World Wide Web, and the Information Highway or Superhighway
are. Much of the confusion is well founded.

The Information Highway, whether or not super, is the least
well-defined of these terms. It seems to have been first used in
1990 by then Senator Al Gore but exactly to what it refers is
unclear. Mr. Gore's father, a U.S. senator, had been instrumental
in establishing the Interstate Highway System in the United States
and the younger Mr. Gore seems to have had in mind another sort
of highway. The metaphor seems valid. The concrete highway
carries motor vehicles and the other carries electronic messages.
But the motor vehicle highway is quite explicit—you're either on
it or not. What, exactly, comprises the information highway?

Some people equate the Highway with the Internet which is not
really correct. The best explanation yet is that the information
highway, sometimes called the *information infrastructure*, consists
of all the transmission systems that carry electronic messages: the
telephone system, all forms of radio, broadcast television, cable

television, and communication satellites.[2] The term could also be used to include the print media: newspapers, books, or magazines, or the conventional means of distributing them, often over a concrete highway or railway.

If the information highway carries the totality of electronic messages, whether they go by telephone, telegraph, radio, or television, then the Internet is a part of it but not the whole thing. The Internet primarily provides a means of communicating among computers. As a transmission system, like the communication satellite, it is not really new: it is a new way of using existing capabilities. Most of us, when we began trying to have our computer reach another computer, did so through the conventional telephone system. Some of us are now able to use the television cable, new high-speed telephone service, or networks of microwave or fiber optic links, whether terrestrial or satellite, to gain a higher bandwidth.

Today, we can send electronic mail, participate in electronic commerce, or search for information among millions of documents. More recently we have been able to pick up some recorded music, radio or TV programs, or motion pictures, or carry on telephone conversations on the Internet. These later applications are in a sense still experimental because there is not yet a widespread habit of using the Internet this way. They could have a big impact on today's broadcasters and receiving equipment manufacturers if they become highly popular. Now, these transmissions are largely free. They are likely to remain so only until their popularity indicates that broadcasters could profit by imposing a charge, bring in enough advertising revenue to provide an acceptable profit, or threaten profit levels for existing media. Napster was an example of the last of these possibilities, taken to court for alleged copyright violations. We'll discuss the tendency toward combining the electronic media in the next chapter.

Technicalities[3]

At the most basic level when one computer talks to another, it is something like two telegraph instruments communicating with each other. Just as an operator must tell the telegraph what codes to send, someone must provide the computer with the messages to send, although today computers are sometimes smart enough to generate their own messages. Most commonly, the sending computer is connected to the telephone network as is the computer at the other end of the line. In between may be a large network of interconnected computers that serve, like the telephone switching centers, only to relay messages. What connects the computer and the telephone, or serves as the interface between them, is a device called a *modulator-demodulator* or, for short, a *modem*.

A modulator is a device for transforming a signal. In this case, it transforms the internal computer codes, which are something like the Baudot codes of telegraphy, described in chapter 7. It transforms them into sounds or the electrical equivalent of sounds. At the receiving end, the demodulating action reverses the transformation. The modem does both jobs. If you are an Internet user, you may have heard the sound of your computer "dialing" the telephone. Once the connection is made you may hear the sounds of the computers exchanging recognition signals. These are high pitched sounds, meaningless to the human listener. It is almost like being connected on the voice telephone where there is an exchange of messages like:

[*Dial tone*]   (The network is ready to receive your call.)

Caller: (Presses buttons to identify intended destination telephone.)

[*Sound of*   (Confirmation to caller that destination *telephone*
ringing]   has been contacted.)

[*Ringing at*   (Signal at destination end that a message is
  destination]   coming.)

Recipient: "Hello." (Verbally establishes that a connection
                has been made to a *person*.)

At this point you can transmit, receive, or simply wait; the line and
all intermediate links are available for your use. The setup continues:

Caller: "This is ___."   (identifies self.)
             "Is __ in?"   (Caller designates the intend-
                                     ed destination of the mes-
                                     sage.)

When talking you are connected to a network of switching
stations. Similarly, your computer talks first to the equivalent of a
local exchange. You tell (your computer tells) the exchange to
whom (which computer) you really want to talk. You may be in
Moscow, Idaho, and your destination is in Moscow, Russia. There
are going to be a lot of intervening computers but you will not be
aware of them. If you are sending electronic mail, one of the
computers will hold the message until your intended recipient calls
in for his or her mail.

When communicating through a voice telephone network, all
the various links among exchanges are set up between you and
your destination and held until you terminate the call. When it is
computer-to-computer, the originating computer connects to an
internet service provider (ISP). The ISP computer transmits your
messages, possibly one line at a time, through the network to the
destination. A message coming back goes through a similar process
to reach you. During any interval when you are not actually
transmitting or receiving, even for a fraction of a second, the
transmitting facilities are not reserved for or charged to you, a
considerable saving in transmission costs.

Messages in either direction may be broken into small pieces
called *packets* and each packet is sent as a separate message. In

to whatever switching center may be available at the moment and in the general direction desired, as shown in figure 12.1. Some modern networks now determine the entire path for a packet all at once.

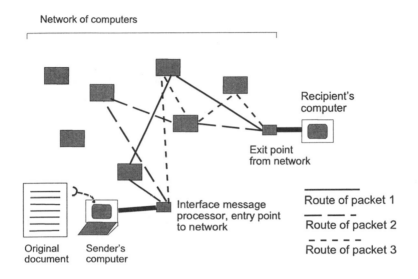

Network of computers

Recipient's computer

Exit point from network

Interface message processor, entry point to network

Route of packet 1

Route of packet 2

Route of packet 3

Original document

Sender's computer

Figure 12.1. *Packet switching.* As a message is keyed into the computer (lower left), it is broken into small units, typically around 100-200 characters and each is transmitted as a separate message, possibly going by different routes depending on the procedure adopted by a particular network. An interface message processor (IMP) at one end determines the route and another IMP at the other end assembles received fragments and presents them in the order sent.

The reasons why packet switching works so well is that you only pay for network time actually used, not for time spent thinking or reading, and when you are transmitting, messages are multiplexed. The result is to greatly reduce the cost of transmission and hence of use of the Internet. Just as a communications satellite virtually eliminates ground distance as a factor in total communications cost, so also does packet switching. There is no perceptible delay as fragments of a message are readied for transmission and none at the receiving end between packets. It all happens too fast.

When packet switched networks first came along, a user would deal with one network at a time, although it was possible to choose which one to use at any time. That meant that the destination computer had to be connected to the network of choice or that the originating and destination nets could link up. Within the United States it was reasonable to assume any given target could be reached, but you could not necessarily reach any computer, anywhere in the world.

The Internet is a means of communicating among networks. It is actually a protocol for networks to observe when connecting among themselves. *This is important.* The Internet is not a transmission system, not a network. It is a procedure for interconnecting networks. Telephone systems have something like this. The different countries of the world have separate systems but are able to communicate by observing common interconnect standards. There is no director of worldwide telephoning; there are only the interchange standards and agreements among separate systems. As pointed out earlier, television has no provision for interconnect standards between incompatible systems. One of the most interesting aspects of the Internet is the lack of a single organization that runs it. There is only a standard and an organization that sets standards. No enforcement. There is no Internet Corporation selling its shares. When you hear about glamour Internet stocks, these are companies that sell hardware, software used to get on the Internet, or companies doing business on it, not the Internet, itself.

Cost

A substantial number of users of the Internet, exact amount unknown, use it through their school or place of employment. Hence, the number of computers in use with modems does not tell us how many actual users there are. For these institutional users, Internet access appears to be free. If you must pay, as from home, an Internet connection will cost anywhere from about ten to thirty dollars per month which should get you at least 100 hours a month and sometimes even unlimited use. The prices vary considerably among ISPs. For example, telephone-based connections have gone down in cost but TV cable operators have begun offering the service usually at a higher price but with much higher transmission speed. They also eliminate the need to dial up the ISP. The telephone companies have countered with high-speed service in some areas using what they call *digital subscriber lines* (DSL) that compact signals, giving the equivalent of more bandwidth. Of late, satellite operators are also offering a similar service. Competition tends to bring down prices.

History

Communication among computers is older than it may seem, given the relatively sudden popularity of the Internet. In the early 1950s computer data were often stored on punched cards (figure 12.2) in which the pattern of holes served as the equivalent of coding with dots and dashes or Baudot codes. These cards could be punched or read by a computer. They could also be read by a machine that could be attached to a telephone line and the contents transmitted to a distant location where the information was punched into a new set of cards. While these machines were not in widespread use, they were readily available from IBM, that is, they were common enough to be a regular, if not best-selling, commercial product. These were extremely slow by today's standards, but if the goal were to send payroll figures for 1000 employees or reordering information on 500 items for a warehouse, they were good enough. Such machines would not have been able to keep up with the rate of flow of data on New York Stock Exchange transactions. But the idea of communications among computers had been established.

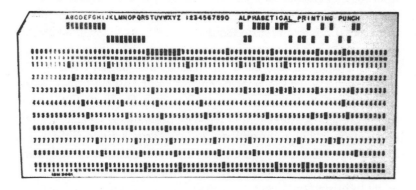

Figure 12.2. *A tabulating-machine card.* Such cards were used for many years to feed information into early computers. At any of the numbered positions a hole could be punched. The pattern of holes in a column constituted a code for a letter, numeral, or punctuation sign. The coding is similar to that of Baudot shown in figure 7.4.

Air Defense and Airline Reservations[4]

In the mid-1950s the U.S. Air Force undertook to build a ground-breaking system to serve the air defense needs of the day. It was called by the odd name of *Semi-Automatic Ground Environment*, or SAGE. It consisted of a network of computers and a ring of radars covering most of the North American continent. The system was also connected to ground-to-air missile sites and to the command centers for interceptor aircraft. When radar detected a flying object entering North America, the computers would track it, display the track to human controllers, and correlate with known flight plans of military or commercial aircraft. The computers did *not* make decisions about intercepting. These were made only by air force officers. But the computers would relay information to the appropriate people and record and transmit their decisions.

Communication between the human controllers and the computers was done through consoles consisting of a keyboard, a display screen, and a *light pen*, a precursor to today's mouse, a means of pointing to something on the screen to ask for more information. Such a console is illustrated in figure 12.3.

To my knowledge, SAGE never intercepted anything but it was a great technological accomplishment. The overall system design was done by the Massachusetts Institute of Technology (MIT). The computers were built by IBM which, not long after finishing SAGE, built a system called SABRE for American Airlines.

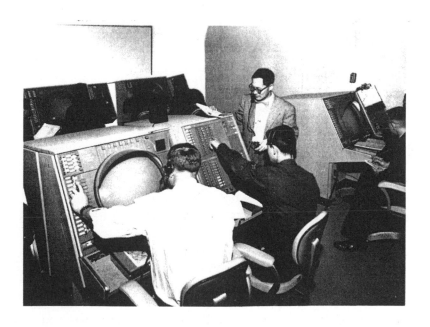

Figure 12.3. *A console used in SAGE.* This workstation was a desk containing a display tube, a keyboard, and a light pen. With it an air force officer could follow aircraft and, if necessary, order an intercept. This rather modern system was operational in the 1950s. Reprinted with permission of MIT Lincoln Laboratory, Lexington, Massachusetts.

SABRE was a reservation system that kept track of all seats on all the airline's flights and handled such other complex problems as crew scheduling. As soon as a seat was sold anywhere, on any flight, the records had to be changed so that the next inquiry about seat availability would reflect the recent sale and not allow a duplicate reservation. A descendant system, still called SABRE, is now the basis for a separate business and is used by many airlines and travel agents. SABRE then branched out into a consumer service called *Travelocity*. Altogether, SABRE is said to make more money for AMR Corporation, the owner of it and American Airlines, than do the airline operations.

Sharing Computer Resources[5]

Until about the mid-1960s computers worked by someone loading in a program and letting that program run, or execute its commands, until it came to a stop. Then, another program could be loaded in. But now, once again with MIT in the lead, a new mode was invented called *time-sharing*. Just like its modern namesake, the time-shared condominium, this meant that more than one program could have running rights in the computer but each would be run in sequence, allocated only a brief period of time in its turn. Computers by then were so fast that they could execute a small bit of each of, say, a hundred programs for a small fraction of a second at a time. Each user, working at a console something like that of SAGE, would see results of a computation about as quickly as if that one person were the only user of the computer. This was because most programs frequently exchange data between main memory and disk storage and this takes, relatively, a great deal of time. During those "read-write" delays, many computer commands not involving input or output could be executed on behalf of other programs. Also, the discrepancy between the rate at which a computer executes commands and humans think or type is enormous. While one person pondered what to do next, or typed in new commands, many segments of other programs could be run and the other users could begin to prepare for their next steps.

The idea quickly appealed to computer users and, although it may not seem so, it was a more economical way to make use of expensive machinery.

ARPANet[6]

While the Department of Defense was spending great sums on computer science research they, or their researchers on various campuses, were frustrated that a researcher on one campus could not use some new program at another because the respective computers were not compatible. Wouldn't it be nice if a person on one campus could communicate with the computer on another and be, in effect, just another user, even if 4500 km away? This was one of the situations that led to the development of the first general purpose computer network, i.e., not one like SAGE or SABRE which were each devoted to a single purpose and open only to a restricted set of users, but one open to many users for many different uses.

A second situation was that there were now guided missiles all over the country and their control required reliable communications. The mechanics could be handled by the telephone network but what if there were an attack against the United States and some of the phone links were cut? (All this was happening at the height of the Cold War.) The solution seemed to be to create as many redundant links as possible and to create a system for controlling the network so that if any link were cut, messages could simply be routed around the break.

Combining time-sharing with a network led to the first packet-switched network. It was called ARPANet, named after the Defense Department's Advanced Research Projects Agency. Although so many persons were involved it is hard to point to one inventor, one of the key people involved was J. C. R. Licklider (United States, 1915-1990), who had been at MIT and Bolt,

Beranek, and Newman, a company headed by MIT professors and much involved in computer time-sharing. "Lick" had been one of the principal proponents of interactive computing that enabled scientists to work directly with a computer, trying out an approach to a problem, seeing computational results almost immediately, changing what may have needed changing, and continuing in this manner until all was well. Time-sharing was a requirement for this. Licklider became the first head of ARPA's Information Processing Techniques Office (IPTO) and got the ball rolling on ARPANet. He was succeeded by Robert W. Taylor who began the process of designing the network and then by Ivan Sutherland who saw to its creation. Major roles in actual implementation were played by Leonard Kleinrock, at UCLA, and Lawrence Roberts, another ex-MITer at ARPA. It's hard to tell exactly who got what idea, when.

Licklider's early musings on how humans ought to make use of computers became the standard way that we do interact with them. We think, we try something out, we see the results, we make some changes and try again. It may be hard for a modern computer user to visualize that as late as the early 1960s it could take from a few minutes to a full day to get results from a computer and then decide what changes to make.

ARPANet to Internet[7]

Following the success of the ARPANet a host of commercial networks grew up. As noted earlier, each network might have links to a different set of computers, but anyone could call into the network and ask to be connected to a remotely located computer, usually requiring a secret password attesting that the caller was an authorized user of the particular service. There was a cost, usually charged to the remote computer which, if commercial, would in turn charge the caller, something like 900 telephone service today. The caller to a 900 area code number pays a fee most of which goes to the organization called, far higher than the communications cost. Also as noted earlier, before the Internet you could not be assured of reaching *any* computer on *any* network. What was missing was a way to go from the network normally used by our hypothetical person in Moscow, Idaho, to a network in Russia. What was needed was a way to interconnect the networks.

Internet to Nirvana

Once the Internet came into existence it virtually exploded in terms of the number of computers available to it and number of users. Figure 12.4 shows this development. In the figure, a site is a computer, a host is a software system within a computer. Similar growth curves exist for printed books and for telephones in use, but they took centuries or decades, respectively to develop. This curve spans only two decades.

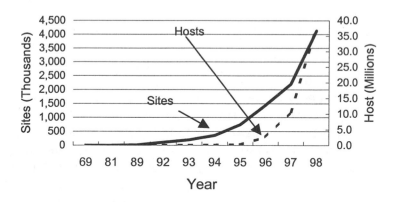

Figure 12.4. *The growth of the Internet.* This shows the number of Internet hosts (software systems) and sites (computers containing one or more hosts) from 1969 to 1996. The sites existing as early as 1969 would have been all experimental. Source: Gromov, History of Internet and WWW.[8]

We now have sites that contain documents, those that provide a service for searching for documents at other sites, those that provide electronic mail service, and those that provide electronic commerce service. We also have sites that offer sound or motion picture recordings, which are forms of documents in our new multimedia world. These can be listened to through high fidelity stereo equipment attached to a computer, viewed on the computer's screen, or downloaded to a disk for later use. A larger video monitor or speaker in another room can be attached to the computer but the computer, not the TV remote unit, is the controlling device (today). In short, it is possible to replace radio and television receivers with the computer and thereby to make the Internet a true multimedia medium. The information superhighway and the Internet may become interchangeable in fact as well as in popular usage.

The following are the major services offered through the Internet.

Electronic mail (e-mail).[9] E-mail is essentially a form of telegraphy but sent from a computer in the originator's home or office and delivered directly to the recipient's computer. The sender does not need to know Morse or any other code. Transmission time is about the same as it was in the nineteenth century, but without the delays of hand delivery. E-mail is a by-product of ARPANet development. Its origin is credited to Ray Tomlinson of Bolt, Beranek, and Newman, the company that was charged with implementation of the ARPANet. In 1971 Tomlinson was working on a method of transferring files but wanted to be able to address an individual person rather than a computer. He invented the usage of an address consisting of a user name or code, followed by @, followed by the designation of the computer with which the addressed person was associated. In large organizations a single computer may serve as the mail station for thousands of individuals.

E-mail seems free to most of us because there is no charge for individual messages. We are charged for time connected to the Internet, regardless of what we do when connected, and many of us do not pay for the connect time. Even if we did, the cost of an e-letter would probably be less than conventional postage. Its use has grown enormously since it first started. Figure 12.5 shows the number of messages sent per year. Prior to 1986 there was little traffic and little counting of messages.

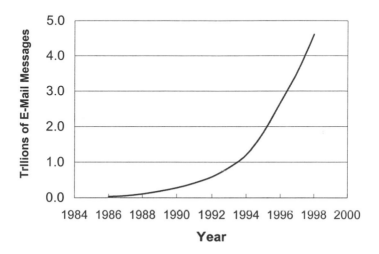

Figure 12.5. *The growth of electronic mail.* Shown are the number of trillions (millions of millions) of messages sent per year. Although the activity started before 1986, the numbers for those early years would not show up on this scale. Source: Wilkofsky Gruen Associates, used with their kind permission.

The great advantages of this method of communication, other than cost, are its ease of use and speed of delivery. A warning: while it is generally looked upon as a private means of communication, there is no sealed envelope. Copies of electronic letters could be stored, hence retrieved, at any number of locations as the letters are transmitted and stored awaiting the recipient to check for incoming mail. Nonetheless, most people who use it seem to consider it one of the great benefits of the Internet.

The World Wide Web (WWW, or "the Web").[10] The Web is a network of documents or portions of documents or programs. It is not a physical network; it is a set of documents that use a system of pointers to indicate where related information might be found. The basic idea is very simple. What were once footnotes or other forms of pointing from one portion of a text to another, for example, to acknowledge a source or suggest relevant additional reading, are now the Internet addresses of the cited documents in the text. The idea of such a world of interconnected documents is due to **Theodor Nelson**. The implementation on the Internet is due to **Tim Berners-Lee**, a British physicist working for the Centre européene de recherche nucléair (**CERN**) in Geneva. The original purpose was to speed dissemination of information on physics research but it quickly spread to other uses. The combination of this form of indexing allows a Web user to simply point to a reference, click the mouse, and in a few seconds see the referenced document. This is extremely useful to any research person who must see the related works. It is also useful in advertising. "See our catalog. Click here." "Place your order now. Click here." Or in the sports news. You're reading about yesterday's ball game and want to know what **Mark Magwire**'s home run record was last year. Click. There's his lifetime record on display.

By way of example, the various names shown in boldface type in the preceding paragraph, while not actual Web addresses, show how the Web is used. If this book were on the Web, more information on any of the persons or organizations named could be found by clicking on the highlighted name (or, actually, the address of a document containing the information). That will result in the indicated document being retrieved and displayed.

Another form of personal communication is the personal Web site. This is a document descriptive of an individual, available to the public. It is used by professional athletes, entertainers, authors, and the like to advertise and perhaps to avoid excessive e-mail.

Like e-mail, the Web may have started out with a specific application and limited user group in mind, but it quickly found nearly universal interest. Computer programs are, in effect, documents and the user can switch from program to program as they do from document to document. A significant problem today is that there are too many documents, including still and motion pictures and sound recordings. There are so many that it is not uncommon to ask a Web searching program (or *search engine*) what seems to be a specific question and yet be told there are over a million documents on the Web meeting your specifications. To counter this, search engines are getting more sophisticated and better able to interpret your questions. They can rank retrieved documents in order of probable interest to you. But the sheer number of documents seems to keep ahead of the intelligence shown by the search engines.

The Web and the Internet are often confused with each other. Remember that the Internet is a procedure for connecting networks, although in popular usage it *is* the network. The Web is a set of documents found on computers connected into the Internet, if we may use the incorrect sense of that word. What connects computers on the Internet are telecommunication facilities. What connects documents on the Web are pointers, addresses, or the equivalent of footnotes. These are not physical links. Documents are no longer just texts or illustrated texts. They may contain interactive programs, allowing users to provide input and able to act upon that input to decide what to do next, as in the simple case of allowing a user to order a book, then acknowledging the order, accepting and verifying a credit card number, and initiating the shipment.

Electronic commerce (e-commerce).[10] A combination of e-mail and searching, e-commerce provides a means of buying and selling by passing messages back and forth over the Internet. In one sense, it is nothing new. The idea of shopping remotely from a catalog originated with Aaron Montgomery Ward in 1872. He had been an itinerant salesman who tired of so much travel and began mailing out catalogs from which customers could mail him orders for merchandise, which he would have delivered by mail or express. His idea took hold and he became very rich, like some late twentieth-century innovators in business communication. Eventually, Ward was overtaken by Sears Roebuck, lasted a long time even with competition, but succumbed in 2000. But Ward originated the idea that, for many, the convenience of shopping by catalog more than made up for the fact that the customer could not actually see the merchandise in advance nor talk to a salesperson about it. Note that this was a multimedia business from the beginning. Multimedia is not solely an electronic phenomenon. Ward needed the *mail* to send out his *printed* catalogs and to receive customers' orders and he needed the mail or an *express service* to deliver the goods. Delivery, in turn, required adequate *vehicles* and *roads*. He also needed an economical *printing service* to produce all those catalogs, although the original consisted of only one page. All the italicized words indicate some form of communication medium involved in such commerce. When several prominent e-commerce firms found themselves unable to handle the load of business around Christmas 1999 it was to a large extent due to inventory control and physical delivery problems, not to electronic communications. But valuable lessons were learned. E-commerce requires more than simply a Web site.

Long before the Internet many catalog sales companies were relying on telephone calls in addition to postal mail to bring in orders. Now, we can have our choice of far more information and illustrations on the Internet than a printed catalog can show.

There are four basic differences between e-commerce and traditional mail-order shopping. The first is the *high speed of transactions*, once a decision has been made to buy something; also possible by telephone. Once a person decides to buy something the order can be sent and acknowledged and payment made by providing a credit card number, whose validity can be checked while the buyer is still online. A second difference is the *amount of interactivity* between buyer and seller. This, too, is possible by telephone. The third is the *amount of information* that can be exchanged. This is new. The Internet can offer far more than a printed catalog. The fourth difference is in the *speed at which content of a Web site can be changed*. This is different from transaction speed. It relates to the ability to change the catalog or description of merchandise offered for sale, or the conditions of sale or delivery, on a moment's notice. Obviously, this cannot be done with printed documents.

People like interactive systems, systems that respond to them, to show another view of that sweater, to show a variety of colors of a product. A printed catalog, however well designed, just sits there. But Lands' End, among the largest mail-order clothing suppliers in the United States, found that the printed catalog stimulated telephone and e-commerce sales. The catalog contains only the information that is obviously on view but it is what initially attracts the buyer. A user cannot ask the catalog for more information but can through an e-commerce system. With modern Web sites users can ask for additional views of the product, additional information, details of shipping dates, and so on. It is fully possible—and is being done—to have the e-commerce buyer talk with a live salesperson or customer service representative. Lands' End can also display a figure with the dimensions of the caller (who obviously had to supply the data) clad in whatever garment is under consideration, to see how a given size or color will appear on the electronic mannequin. This is not quite like trying it on in the fitting room, but a giant step ahead of pure catalog buying. It adds up to a totally different experience, or it can.

Unhappily, e-commerce can also be done badly. It becomes easier for a seller to hide information and disappear, in the sense of no further customer contact, after a quick sale. It is also critical that, however fast the order processing, there must be a delivery system that can bring the merchandise to the customer in a short time. A television commercial, frequently run in 2000, shows a woman trying to buy something in a store only to be told it is not available. She says she saw it advertised on that company's web site. The clerks say it has nothing to do with them. I have had the very same experience in real life with a bookstore chain. Their e-commerce business was a separate corporation from the bookstore chain, although both have a common corporate owner and similar names. The people in the store would not help make an on-line order, could even charge a different price for the same book. The e-commerce company made it very difficult to talk to a human if there was any sort of problem. The result was very frustrating for customers. Why is this written in past tense? The company is gone, swallowed by a competitor. These problems are not the fault of the Internet, just as poor programming is not the fault of television technology. People will have to learn how to use these facilities in attractive ways and customers will have to learn to demand better service.

E-commerce is often referred to as virtual shopping or the vendor as a virtual store. It is not virtual. It is real. It is as real as Montgomery Ward or Sears Roebuck were in catalog sales. Amazon.com is a real seller of real books. The catalog, whether printed or electronic, is a surrogate for the goods, enabling the buyer to make a decision and place an order without ever seeing the actual object. There really is a storehouse full of merchandise. What is different is how the buyer sees the merchandise and places an order. E-commerce has become very popular. It saves time and most customers are clearly finding they can trust most of the e-stores.

Impact

There are already people calling the Internet the medium of greatest impact ever, or certainly ever since Gutenberg. Surely, in terms of number of people undertaking to make use of a medium over a given period of time, it is undoubtedly the world record holder. Is it changing society? It seems to be. Is it for the better? We are not sure yet.

Have there been changes or will there be? Let me count the ways: serious research (including on the stock market), personal communication, commerce, education, globalization, and social cohesion.

Research

To the serious person engaged in research, the Internet, and especially the Web, is a great boon. This is true whether the person is a scientist, student, stock-market trader, lawyer, or newspaper reporter. Starting with any document pertaining to the desired subject, the researcher can quickly jump from one to another based on clues in the first, until the desired information is found or the searcher feels all possibilities have been exhausted. (My own experience has always been that, when frustrated, you call a knowledgeable friend.) Unless all the documents scanned were available in the local library, it could take days or weeks to track down all of them and hope they were not in circulation when needed. Even if they were all in the same building it could take many hours to find first the book, then the part you need.

It is not just a matter of saving time. There is a qualitative difference in doing research this way. Everything is fresh in your mind as you do the search and if it becomes necessary to change what you are looking for, you can do it without feeling that you have to start a long process entirely from scratch.

The down side of searching on the Web includes the aforementioned sometimes excessive number of "hits" that can result from even a specific question and the tendency to assume that "everything" is on the Web, is authoritative, and is available free. None of these assumptions is true.

Personal Communication

Some decry the dying of the art of letter writing. They may well be right that it is a dying art. But media change and people change with them. There was a time before letter writing and we may be headed for a time after it. Some decry the volume of e-mail that accumulates daily, much of it of no interest. But this is really no different from an excess of postal mail or telephoned messages that some of us are afflicted with. Some years ago, a colleague and I were to interview an assistant secretary in the U.S. Department of Transportation. He was running late and while waiting we chatted with his secretary who showed us a stack of pink telephone message forms, around half an inch thick, that had accumulated since the man in question had left the office earlier that day. It seemed obvious that returning every call would easily have taken half a day. By then, more messages would have arrived. One of the secretary's great strengths was her skill at determining which calls could be passed off to someone else, which might be ignored, and which called for a quick answer. The overload problem existed before e-mail and will probably persist indefinitely. There is software that can filter mail but like Web-searching software, it is imperfect. Don't blame the message service if there are too many messages.

Commerce

What we have in e-commerce is a faster and more convenient way to shop. Some retail stores, following the downsizing trend now so popular, offer less in-person service than they once did, which has the effect of making catalog shopping more attractive. If you can't get in-person service anyway, you might as well stay at home and shop.

On the other hand, one of the most glamorous of the e-commerce firms when the idea got started, Amazon.com Inc., does not make a net profit (as of end of 2000). This, of course, may change in time. Will all the new glamour companies be in business a year from now? How will they fare in the face of competition, most having started as the only firms doing this particular type of business? What will happen to the start-ups when established, rich businesses decide it is time to enter this market? What would be the effect on stores and shopping centers and the people they employ if this were to become the primary means of retail selling? Shopping is a social experience as well as a commercial one. How will the lives of shoppers change? Governments have been reluctant to regulate or tax e-commerce, apparently to avoid squelching a new and still tender industry. But what if e-sales cut heavily into conventional sales? They would also cut into sales and real estate taxes on the old-fashioned stores.

Education

There are several ways that Internet technology can assist education: help in creation and presentation of course materials, the modern-day equivalent of books; encouraging students to find information on their own, evaluate it, and use it to create their own reports or interpretations; word processing; and simply being able to correspond with potentially helpful people all over the world.

The development and presentation of learning materials and word processing are really not particularly Internet-related, but well-designed materials do help with learning and motivation and the Internet becomes a means of distributing information at low cost.

When students dig out information for a project or a course by themselves, they are likely to become more involved with it, perhaps go a step or two farther than required and to retain the knowledge longer. It seems to be the fun of the quest and, of course, having some success with it, that provides the motivation, far more than reading dull textbooks. If there is a down side to this it is that it becomes tantalizingly easy to find some material for a project and incorporate it, verbatim, into one's own work. That is not only morally wrong, it involves little learning and may have a negative effect on attitudes because a student who has done no real work or learning may earn a high grade using someone else's work.[12]

I personally feel the role of computers in general in the class room has been over-emphasized. Not by any means are they unimportant, but it is far more important that students master critical reading, looking, and hearing, and oral, written, and graphic expression of thoughts. If they can do these and make use of a computer, so much the better. But learning to use machinery does not replace learning to use language, music, gesture, and art.

Another aspect of education is the possibility of delivering instruction through the Internet. To many this is the solution to ever increasing costs, particularly of higher education. The down side is the loss of personal contact between teachers and students and among students. From the lower elementary grades through graduate school, this interaction between teacher and student has traditionally been important. There is some merit in relieving students of the need to be in a particular place at a particular time, allowing them to e-mail questions to teachers and get answers quickly, and hold e-mailed discussions among themselves. Still, many educators worry about the tendency to want to switch to this mode because it seems the new, innovative thing to do. Will it work well? It may take a long time to find out.

Globalization

One of the most frequent comments made about the Internet is that it is bringing us the global village, the term popularized, but not really defined, by Marshall McLuhan. Like many computer users, I correspond with people all over the world. This global aspect of e-mail is, to me, one of the greatest benefits of the Internet.

My correspondents are either in the category of friends or relatives or they are professional colleagues. They are almost always people I already know, who have been suggested to me by someone I know, or who are responding to something I have published. The point is that some kind of link exists before we contact each other. But a village is a place where just about everyone knows everyone else, whether or not they all work or socialize together or like each other. The global Internet gives us global affinity groups, not a village in the old sense. We do not *have* to meet people we feel no affinity for on the Internet, but we have to in the real village. We may be building cocoons for ourselves or we may be opening up the world to communication across all sorts of lines, physical as well as social.

Social Cohesion

The great attraction of globalization is that it pulls people together who are geographically separated. However, there is some danger of widening the gap between the affluent, educated population and lower income, less educated segments of society. Not everyone has a computer at home or Internet access where they work, certainly not minimum-wage workers. Those without resources are not learning how to use the tools that are becoming evermore required for any well-paying job or challenging school. The more advanced technology becomes an essential part of education and job performance, the more we may be widening this gap, pushing away members of our own society while we reach out to people like ourselves in other lands.

If we try to maintain the perspective of history, we would have to say that the Internet is too new for us to be able to gauge its effect on society. But it has hit with such stunning impact that we can no longer decide to wait to see if it will be a flash in the pan. It has captured users in the home, in the schools, and on the job. It has already changed the way many businesses and educational institutions operate. It may radically change the way entertainment and news are delivered to us. It's here; it's important; it's not going to fade away. It could easily become as important as printing.

Notes

1. Berners-Lee *Weaving the Web* and "The Semantic Web"; Hughes, *Rescuing Prometheus*; Naughton, *Brief History*. Lewis, *The New New Thing* describes the development of Netscape, only a part of the overall story, but fascinating.
2. United States General Accounting Office. *Information Superhighway*.
3. Schatt and Fox, *Voice/Data Telecommunications*; Sherman, *Data Communication*; Stallings, *Business Data Communications*.

4. Hughes, *Rescuing Prometheus, 18-67*; Everett, Zracket & Bennington, "SAGE."
5. Pyke, "Time Sharing." Published in 1967, but since then time-sharing has been relatively standard in large computers.
6. ARPANet. Hughes, *Rescuing Prometheus,* 255-300; Licklider, J. C. R. "Man-Machine Symbiosis."
7. Abbate, *Inventing the Internet*; Rowland, *Spirit*, 295-337; "The Internet," *Scientific American*; McKnight and Bailey, *Internet Economics*; Naughton, *Brief History*.
8. Gromov, Gregory R. "History of the Internet and WWW."
9. Specter, "Your Mail."
10. Berners-Lee, *Weaving the Web* and "The Semantic Web."
11. Gladwell, "Clicks"; Munro, "The Ongoing"; Drucker, "Beyond"; Westland and Clark, *Global Electronic Commerce*.
12. Meadow, *Ink*, 148.

Part 5

Looking Backward and Forward

A Spanish proverb I once heard states that if you want to make God laugh, tell Him your plans. To this we might add that to make Him *really* laugh, tell him your predictions. Predicting the future of technology is a daunting task. Most people who try it, or at least who try to go beyond just a few years, miss terribly. Commonly, these predictions are based on extending what we have now. They cannot take into account what has not yet been invented. Also, there is the matter of acceptance—which technology will succeed depends not only on what is invented but what is accepted by the using public. Public acceptance depends on a number of factors, such as the taste, needs, and wants of individuals, or the desire to be a leader (or follower) in adopting new ideas and tools. Perceived needs and wants may reflect a person's own personality or have been formed by advertising or other forms of social influence. We're going to briefly review the communication technologies of the past that had major impact and explore not what *will* happen in the future but what *might* happen, what it seems reasonable to assume. Whatever is predicted, by whomever, we'll be surprised by the reality.

In spite of all the changes, certain basic principles of communication remain and we shall look at these in light of the many changes since the early days of cave drawings.

We conclude the book with a chronological list of one hundred notable dates of occurrences in the history of communication.

13

Summation and Projection[1]

Introduction

We are communicating creatures. Other animals are, too, but none are like us in the variety of media we use and the range of content we express. One consequence of our penchant for communication is that every major change in communication technology tends to bring with it behavioral changes in individuals and society as a whole. That sentence needs clarification. What is a "major" change in communication technology? It might well be that a major technological change is one that causes a major societal change. If so, we're involved in circular reasoning. Is there an objective measure of importance, aside from societal change? There can be variations in what we mean by societal change. Perhaps money earned, investment attracted, jobs created, or number of people who adopt the new invention could serve. But importance derives from effect, not anything inherent in the technology.

Technologies That Accompanied Major Change

We cannot list all the inventions that were never accepted. Most we have never heard of. Some were adopted and then, for a range of possible reasons, dropped. While not communications technology, the chemicals DDT and thalidomide are examples of developments that were initially accepted for their seeming benefits, then rejected because of unforeseen side effects. Some are still controversial, such as nuclear reactors for power generation and genetically modified food. Some have never yet arrived, such as controlled nuclear fusion, a "clean" source of the amount of power modern society needs, and so far, fully accepted, readily available, nonpolluting road vehicle engines or fuels.

Some inventions came and went: eight-track musical recordings, betamax television recording, and the picture phone method of telephoning. The last added a viewing screen to the telephone in the 1980s but it never caught on with the public. Betamax, as noted in chapter 10 is still considered by some to be superior to VHS, but VHS won a battle for recognition among the companies that had originally been partners in the development of betamax. Eight-track cassettes were just not needed after the smaller two-track cassette recordings proved far more popular. Many predicted that computer-assisted instruction, dating back to the 1960s as an invention, would revolutionize education. Very simply, it has not.

Microfilm is another medium with a checkered history. Its use dates back to the Franco-Prussian War of 1870. When used for archiving newspapers or legal records, for many years there was no alternative to microfilm until computer memories got big enough and cheap enough to be used for archival purposes. Still, retroactively converting old historical records to computer images is slow and expensive; so mostly only the new ones are so treated. But before computers could handle such volumes of information, microfilm seemed the only way to handle large masses of text and pictures. Indeed, an article by Vannevar Bush, science advisor to President Roosevelt during World War II, proposed a new way to handle the search for information.[2] It was to be a microfilm machine with advanced logical capabilities, some not yet reached by modern computers. For all that, microfilm was never popular with those who had to use it. It is cumbersome, hard on the eyes, slow in searching and unable to search on the content of photo images of text. So, when the computer finally was able to compete with it, microfilm tends to lose decisively.

We can test how a particular machine works mechanically before bringing it to market, but we cannot do much of a job of testing public acceptance and effect on society beforehand because these may take a long time to develop.

We don't really know much about what our ancestors were like before speech and gesture. It's hard to imagine a race like ourselves whose members could not even grunt or gesture in meaningful ways to each other. These faculties must have been with us from the time we became human. Either music or drawing was the first consciously developed communications technology. Writing evolved from drawing and sculpting. It helped in government, commerce, agriculture, and engineering. It originated in a society that was fairly advanced for its time in these activities, whether we grant priority of invention to Sumer, Egypt, or China. These important activities and writing then further developed together. Innis, McLuhan, Logan, and Shlain pointed out that the alphabet, linear writing, caused linear thinking and led to the rationalist-scientific thinking so characteristic of the western world That, certainly, was a major effect. We still tend to think that way, by and large. Regardless of whether the alphabet changed society or a changing society created an alphabet to further its progress, the alphabet signaled great change.

Printing with movable type is almost universally acclaimed as one of the highest, if not the highest impact of communication inventions. It came to a European society already showing the stirrings of change in learning, art, and religion. Again, it does not really matter which came first. The social change and the technological change supported each other's development.

Around the time of Gutenberg, serious ocean exploration began out of Europe. This required ships capable of making a transoceanic voyage and the navigational instruments to guide them.

Neither the ships nor the instruments were purely communications devices, but they were essential in communication among diverse peoples. The ships carried people and goods, bringing European culture to a large part of the world, and great wealth and some new ideas into Europe. True, the communication was not necessarily to the benefit of many of the non-European peoples, but for good or ill it did have an effect on the world.

It was a long wait from Gutenberg to the next big telecommunications change, the electric telegraph. This was the first major industrial use of electricity and it almost immediately became important as an alternative to carrying the mail. It quickly affected institutional users but did not much affect the average individual. An operator had to be highly skilled. Its use required wires connecting one station with another. To send a telegram required that the message be delivered to the telegraph office and to receive one meant that the message had to be delivered to the recipient. This meant hand, not electrical, delivery. The message might go thousands of kilometers by wire in an instant, but it could take hours to be delivered at either end. What telegraphy did was enable railroads, other businesses, and armies to be operated with greater speed and efficiency, orders to be sent and acknowledged, and news to be quickly disseminated. It did not, of course, move the commercial goods.

The telephone, adding tonal variation to the telegraph, caught on quickly with a great number of users. This time the instrument came into the home and, because it could be used by almost anyone, the cost of wiring, included in the price of service, was worthwhile for customers. The telephone is interactive. There is immediate response to speaking into it. Not only did people take to it quickly but it enabled several changes in the way they lived. Progress in construction techniques and invention of the electric-powered elevator made it feasible to have high apartment buildings in the city whose residents were still connected with the ground for casual communication and safety. Together with the automobile and electric streetcar, the telephone also enabled those who did not want to move *up* to move *out* into burgeoning suburbs. Again, being distant from the city center did not mean being out of touch with it.

Back to the notion of interactivity of the telephone. Today, we find ourselves more and more talking to and being talked to by machines. This is still interactive in a way but it gives one pause. Will we ever want to reverse that trend, return to a situation where if a telephone is answered, we know we will almost always get some actual person on the line? The point is that the automated systems are often woefully inadequate in helping the caller find the appropriate person to talk to.

Radio came in two steps. The first was wireless telegraphy which, like the wired telegraph, was only directly used by organizations: shipping companies, governments, and commercial houses. Probably its major early impact was at sea. *Titanic* aside, radio saved a great many lives at sea and enabled passengers to stay in contact with their businesses and social lives ashore.

Although there had been some experiments with broadcasting by telephone, the ability to transmit by radio led to the first real broadcasts. Broadcast radio is not interactive, but it was very attractive and made listening to it a family thing and something to talk about with friends the next day. On-the-spot news reporting had started, in a small way, with Marconi's wireless yacht race reports. Broadcast radio gave us the *Hindenburg* crash report, political speeches, drama, sports, and war reports. Still, radio is cool in the McLuhan sense. It requires the listener to fill in with imagination what cannot be presented explicitly.

Television began to change the coolness of broadcasting. While initially anything but hot, it gradually improved its color, resolution, and camera technique to the point where we can now quite easily give ourselves up to watching, letting it do the work of the imagination. An early attempt at interactive television went nowhere but it has become somewhat interactive in an unexpected way. The remote tuning device, in the hands of a virtuoso user, enables the watcher to create programs of his (it seems to be largely a male phenomenon) own choosing. It's not necessary to sit through a commercial during a ball game. The channel can be changed back and forth to another game or to news. A fast hand can keep two or three programs going in this way. It provides an experience to the viewer that broadcasters did not intend and probably do not like because the interruptions could affect audience appeal and because channel switching tends to come during commercials. The remote control provides its user with a virtual broadcast. True interactive television is on the way, when viewers can talk back to the producers, whether to register opinions on the current program, order merchandise, or vote in an election.

The Internet came to us in the 1990s, although its antecedents had been in limited use for two decades before that. Not a fundamentally new means of transmission, it is a new way of using existing facilities and is attracting a great deal of user attention. The Internet is very much interactive. Feedback is or can be immediate. We get video and audio as well as print-like images. Because it is now capable of carrying radio and television programs as well as serving as a voice telephone system, its ultimate impact is hard to predict. How will these different transmission media interact with each other? Which carriers and broadcasters will survive? Let us now take our hesitant look into the future.

The Coming Changes

We will look forward in three ways: (1) at immediate, basic changes in telecommunications technology; (2) at systemic changes occurring over the whole field of telecommunications, not restricted to a single technology; and (3) at the idea of limits to growth, essentially whether the tremendous acceleration of the past few decades can continue and what the effect might be if it did.

Basic Changes

The major, basic changes that are coming have all been mentioned: high definition television, wireless telephone, radio and television broadcasts, and motion pictures via the Internet. Fiber optic cables will continue to proliferate as they offer great bandwidth, freedom from interference from electromagnetic radiation, low rate of attenuation, and lighter materials for the cables. Transmission from the user's location to an Internet service provider will continue to increase in speed by a combination of signal compacting methods, especially over existing telephone lines, digital representation, and use of fiber optic or television cable or satellite transmission facilities. Consideration is even being given to using electric power transmission lines for telecommunication.

Protection of the privacy of communications will probably continue to be an issue. Encryption of mail, commercial transactions, and copyrighted material is possible but controversial. Governments tend to resist complete encryption by private users unless they can have the key because of possible illegal activity. Individuals resist the governments' resistence in order to guard their own privacy. Limitations on content (pornography, hate literature, libel, encroachment on copyrights) will continue to be issues for some time as governments struggle with how and whether to enact and enforce laws to govern the freewheeling Internet. Freedom of information laws enacted by various governments may be expanded, giving individuals rights to more information in the hands of government and private investigative, educational, or health agencies.[3]

We might one day see three-dimensional television by use of holographic projectors and we might smell odors as appendages to dramatic or news presentations or advertising. (We already do transmit odors in some printed magazines—advertisements for perfumes.) Want to show the ocean? Might as well send the smell of salt water as well as a picture and the sound of waves breaking on the shore. How important might these be? They seem almost frivolous now but could add to the sense of reality of what we might some day call *telesensing*. Encasing the person sensing a transmission in a full body suit could enable pressure or resistance to be applied as necessary to make the person feel as if involved in the action being portrayed.

One of the great changes definitely coming, to a much greater extent than now, is not directly visible to the user. That is digitization of transmissions. If a signal is converted from its original analog form into digital, its transmission can be more reliable and faster. We have seen this in music recording. A great deal of what is now in print could be sent as digital transmissions, as we are now seeing in e-mail, news services, and e-commerce on the Internet. Books and magazines could be sent this way, to be printed at the receiving end, since good quality printing is still preferable to electronic display for reading purposes. Electronic paper might encourage this, since it could provide paper-like reproduction but on a reusable surface. Nicholas Negroponte pointed out that bits are easier to transport than atoms.[4] By this he means it is easier and faster to move messages in electronic form than to move printed materials. This is indisputable. Although electronics may not displace printing in the short term, eventually it seems inevitable. The electric telegraph was an early example of exactly this phenomenon. It drove the Pony Express out of business.

The size and speed of computer components keep changing. Size decreases and speed increases, giving us the truly portable computer and cellular telephones. Remember, today's laptops are far faster and more logically powerful than the room-size computers of half a century ago. Many problems that we routinely handle today with computers we could not have undertaken then; the time required would have made it impractical. A simple example is the payroll operation of a large scale employer. It isn't calculating the product of hourly or daily rate and time worked, it is keeping track of complex tax regulations and variable deductions. Another example is fingerprint matching. Now police can do a search based on a print of a single finger. Fifty years ago that would have been impossible. Now we can do a good job of translating the language of documents, critical in such contexts as the United Nations, the European Union, or even bilingual Canada. What will we do a few decades from now that we cannot do today?

There are some really far-out communications possibilities the effects of which we cannot possibly anticipate. We mentioned that there is beginning to be some speculation that the speed of light is not the absolute limit to speeds in this universe. How, if true, might that fact be harnessed to improve telecommunications? There has been speculation by serious scientists that one particle can influence another at a distance, without any known means of communication between them. If true, could this lead to instantaneous transmission over distances without the need for electromagnetic waves? Don't expect this at your neighborhood Radio Shack anytime soon. But, what would that do to us if it did come about? Suppose artificial intelligence also developed rapidly and we could program a computer to detect and *interpret* brain waves coming from a remote person. I don't mean to recognize brain damage as we can do now, but to understand what is being thought? That prospect seems horrible, surely the ultimate form of invasion of privacy. Could we teleport, transmit objects, including animate ones, using telecommunications? They do it routinely in *Star Trek*. That would certainly solve the delivery problems for e-commerce operators but, over all, would it be a better world? All these thoughts need to be considered first by the science fiction writers.

System Changes[5]

Two major system changes are already in progress: *disencumbering ourselves from wire* and *converging of media*. Doing away with all or most wire connections does not merely mean that we can now talk on the telephone as we walk down the street. It means that we need never be out of touch with whomever we want to be in touch with. It means that societies that never built much of a wire-based communications infrastructure do not need to do so in order to use the latest communications equipment. A wireless world implies a great cost saving to transmission services much of which is passed on to consumers. It also can increase the number and economic status of people who can use the telephone. Wireless communication does not mean that all transmissions would go by radio, but that any user, anywhere, can be in radio contact with a facility that can resend the signal over wire or cable.

Convergence means that what once were distinct media—telegraph, telephone, radio, television, fax, computer-to-computer transmissions—are or can all become manifestations of one transmission mode. Each device at the receiving end could detect what is being sent—text, sound, video, odors, bodily pressure—and in what form. Then it can do what needs doing to display or store the information appropriately. The receiver can note that what is coming is a fax transmission. Its content could be displayed on a computer screen, printed, saved on a disk, or presented orally through a voice synthesizer. Another incoming message might be a television program. It could be saved or displayed on a large, high resolution screen, single frames might be printed, or its spoken words could be converted to digital text and stored as such.

All this would be at the choice of the recipient. In the long run, convergence can reduce the number of control devices we have in the home or office, with the computer being the central control and switching center for all the media. The television becomes simply a display screen for the computer and the voice telephone a microphone and speaker for it. Whether we're dealing with wire or over-the-air transmission we would only need a single cable or antenna connection for the building we're in. A sound transmission could be converted into digitized text by use of voice recognition software, limited forms of which are already available. As other, new forms of display are developed, such as HDTV or three-dimensional, holographic displays, we would only need to connect a different display device and perhaps some software but not a whole new receiving mechanism.

Social Change

In the past huge changes have followed the major new communications developments of writing, the alphabet, transoceanic shipping, the railroad, the telegraph, the telephone, radio, and television. What hath the Internet wrought and what wonders will it yet work? *Entertainment.* The manner of delivery of entertainment will change, and when the technology of communication changes the content tends to change, at least to some extent. Radio and television undoubtedly made some inroads on people's available time for reading. How much free time is there left for Internet entertainment to take over? There are indications that Internet use by teenagers is beginning to make serious inroads on their television viewing time but that trend, too, could change. Maybe reading will come back, spurred by the popularity of Harry Potter. There have also been hints that Internet use is beginning to taper off among youngsters. What's next? An example of a relatively recent change is in popular music. Rock and roll seems to need the environment of a large audience to be "properly" presented and broadcast rock and country music seem to need to be acted out as well as sung or played, hence both increasingly depend on the combination of video and audio.

Information availability. The ready availability of information has already made a change and more will come. What e-commerce offers that telephone cannot provide is more product information and rapid changes to files, reflecting inventory status, style changes, shipping delays, and so on. The sharp growth of self-trading stock market services could affect market performance and stability since many an amateur can now trade without a broker's advice. Education will change because school systems seem almost obsessed with computers but how they will change the educational accomplishments of students is as yet unclear.

Most people are not really skillful at searching for and evaluating information. That lack could cause not only education but management to be negatively affected as more people come to depend on the Internet but do not use it effectively. On the other hand, as schools begin to teach the finding, evaluation, and use of information in a serious way, more users will be able to take real advantage of this cornucopia of information and the resulting more knowledgeable society could be a far better one.

Any discussion of what information should be made available brings up questions such as: What is true? What is of high quality? What is appropriate for any audience? Telecommunications people do not answer these questions. They are for the moral leaders of a society to answer. But in the slower paced world of print, certain authors and publishers have become trusted and we users must to some extent rely on past performance to judge the veracity of new information. This notion could carry over to the Internet, but it might not. So far, search engines do not perform the quality of filtering we expect of the better publishers. No other kind of institution has yet stepped forward to offer such a service.

Free flow of information. We don't seem to have to worry about excessive restrictions on information flow. On the contrary, there is reason to worry about the lack of restrictions. One aspect of this is the just-mentioned lack of editorial control, i.e., quality control on what gets published on this new distribution medium. Anything can be published and unwary readers do not know what confidence they can have in the information they find. Lack of editorial control also means that pornography and hate literature are freely available. What this does to a society is not well established, but it can hardly be beneficial.

Lesser known is the effect of the flow of information across international borders. One example of the problem is that some countries do not like to be flooded with entertainment or news from another. Canada, for example, with one-tenth the population of the United States, would like to be able to protect its own entertainment and news media from domination by American media. Other countries may object to what they feel is excessive sex, violence, or disrespect for law in exported print, television, or motion pictures. Control of information flow is physically hard enough when dealing with print and motion pictures shipped in cases. It is virtually impossible with broadcast media or the Internet. To all this add copyright encroachment and libel originating in another country. We will need new coping mechanisms *or* we need to learn how not to rely on others but to make our own judgments on content and legality and act responsibly.

Reaching out and cocooning. The new wireless world enables us to keep in touch, to call for help when the car breaks down, to find out where the roaming teenagers are at the moment, to call for medical help when needed. It also can mean never being out of touch with the job or the stockbroker. So, we can talk to more people or to their machines, be connected to more people than ever before. At the same time we may come into direct contact with fewer and we may have less and less time to relax and maybe think. Many of us today decry a world like this but the question is will the people of the future, living in it, decry it? That is hard to predict.

Globalization. What will be the effect of a growing ability to communicate with people all over the world? Again, hard to predict, but what has already been shown to be true is that electronic media bring information about one country to another. Among the positive effects are that people in unfree countries can be in touch with those in free countries and this can provide a strong incentive for change. One of the first steps a totalitarian leader must take is to restrict the free flow of information. Also, Internet users, anywhere, can learn about other countries, make friends in other places. As well, it is ever harder for a totalitarian country to bar information about what life is like in free countries.

Among the possible negative effects of globalization are: (1) People in poorer countries see visions, particularly of some of the excesses of the industrialized world, that inspire demand for the same kinds of things. Not unreasonable, but if they are simply unavailable in the viewers' countries it can make for considerable unrest. (2) There may be a loss or weakening of a culture that cannot produce the variety and attractiveness of mass media that larger countries can. (3) For better or worse, individual countries might lose whatever controls they may wish to exercise over their own people and what flows across their borders.

Law and regulation. Regulation of communication, in one way or another, has been with us for a very long time. In chapter 4 we mentioned the communication by fire signals from Jerusalem to Babylon that was to occur on the first day of a new lunar month. That communication had to be carefully controlled to assure that the message was truly sent on the first day. In 1710 in England, Queen Anne's Statute was enacted to provide for copyright, although there had been a form of protection earlier, controlled by printers' guilds. But the law was not so much intended to protect authors and publishers as the Crown which, by this law, could control much of what was published. More recently we have the example of the control over allocation of radio (hence also TV) frequencies to avoid complete chaos on the airwaves. But, certain political decisions inevitably are made when limited resources are allocated. In the United States the Federal Communications Commission is largely responsible for electronic communication systems (not the press). But, new changes come quickly and the law and regulation-making processes change slowly.

The Internet is so new, makes transmitting information so easy, and is so technologically different from what we have had in the home before that there is little regulation of it, anywhere. An exception is in the registration and ownership of names of sites. There is a licensing procedure, not government controlled and recently there has been some pirating of names. Once a business establishes itself on the Internet, suddenly finding that someone else is using the domain name that identifies the company is devastating. Potential customers using the most obvious Web site name, may find themselves talking to the wrong company, or even to a criminal organization.

Issues regulators will have to cope with are: (1) *Control of or restrictions on content*, mentioned earlier. It is extremely difficult, some say impossible, to prevent violations of any content restrictions. Generally, control would have to be exercised by after-the-fact punishment. (2) *Cross-border transfer of information*, also mentioned earlier, involves the question of a nation's right to restrict what is imported into its domain. The question arises when protecting fragile local cultures or when "protecting" citizens against foreign ideas. (3) *Growth of communication industries*, typified by the merger of America Online and Time-Warner. Is it healthy for anyone that so much of the news and entertainment industry is in one corporation? (4) *Taxation*. How is taxing to be handled when transactions are conducted through the Internet?

Who gets the sales and income taxes? This, of course, also asks who loses sales and income taxes.

Education. What can be more important to a society than education, the process that determines what its young people will grow up to be? Will education be helped by more of the ready availability of entertainment? Of the availability of overwhelming amounts of information, possibly without corresponding user skills to filter and evaluate it? Will students learn to find, assimilate, and use information or merely to copy it?

Limits to Growth?

The world projected so far is one in which there will be a vast increase in the variety of transmitted signals. As electromagnetic transmission can reach just about anywhere without the need for wires, we are witnessing higher speeds, more variety, and just plain more information or entertainment available through the someday-to-be-merged electronic media.

We have shown the increase in Internet usage at a rate that astounds everyone, and it keeps on growing. But is there a limit? Will the growth someday taper off or even halt? Will the amount of information available simply become overwhelming and hence useless to us? We have reached the point where almost all homes in the United States have telephones, radios, and televisions. We have a way to go yet with computers but the trend is that way. Even so, manufacturers keep finding ways to improve the technology and convincing us we need more and better machines. In that sense, there may be no limit to growth.

Engineers seem always able to find ways to speed transmission or increase bandwidth and to store more and more information on generally accessible computers. But, can we use or do we need all this information? As the amount available increases, do we increase our ability to be selective? If not, we are going to decrease the amount of truly useful information we find because we have just so much time for looking and assimilating. Computer software that helps us find information is improving but although there is no measure of this, my own sense is that we lose ground daily in our struggle to find what we really need in that vast morass of material.

Users of information, whether serious researchers or those in search of entertainment are going to have to learn to become adept at searching for what they want and evaluating what they are offered. How do you find what you really want in a 500-channel television world? How do you find the key information for that term paper for school? How do you find the details about the backgrounds of the board members of a company your company is thinking about acquiring? How can you anticipate the new products your competitors are developing? Answering these questions requires effort and skill, without which we are at the mercy of mechanical selection procedures which, rather than sharpening our wits with more and better information, could dull us with second-rate choices. We have the innate ability to do the searching and to separate the truly useful from the chaff. We have to sharpen our ability and be willing to expend the effort.

Lands' End, one of the largest mail-order clothing firms in the United States, and a leader in e-commerce, still depends on its printed catalog. Customers tend to call (by telephone or Internet) after viewing the catalog. Lands' End continues to distribute catalogs because, today, they are the instrument that creates sales, that motivates customers to write, call, or find Lands' End on the Internet. Perhaps the day will come when these catalogs are no longer needed.

Finally, we must realize that what we transmit with all this high-speed transmission is information and that our capacity to assimilate it is limited. Might there be a law of diminishing returns even for this commodity?[6] E-commerce can take orders and can sell information (e.g., shares in a company) but cannot deliver your new sweater. That has to come by more conventional means, until we reach the futuristic modes, like *Star Trek*'s ability to "beam it up." Until we can use telecommunication for transportation, we remain dependent on such mundane forms of communication as delivery trucks, in turn requiring good roads.

A Review of Principles That Do Not Change

For all the changes we have seen, what does not change or has not changed are the fundamentals.

It is still important to recognize the difference between direct and indirect representation of concepts to be communicated, whether we're drawing pictures on a cave wall or developing icons for use on a computer screen.

It is still important to understand that the context in which a message is received affects its interpretation by the recipient. It is never enough for an originator of a message to say, "I made it perfectly clear that . . . ." It is always necessary to think how clear that message will be to the other party. I often wonder, when I hear political speeches, whether the candidates realize that their two-sentence encapsulations of a new health or social security plan is not clearly understood by anyone. Or is the real message, "I'm saying soothing things to you. Therefore, you should vote for me."

It is still important to recognize the effect of media, hot and cold. What is the long-term effect of too much of hot media on young minds? Hot media do not encourage imaginative thinking.

As new media, or major changes in old, are developed, they do not usually drive out old media. They change them. Granted, the Pony Express was pushed out by the telegraph, but not horse-powered express, at least not for some time. Wire-based telegraphy is just about gone, but Internet-based telegraphy thrives. Television did not drive out radio and, in spite of the insistence of parents and critics, electronic media has not yet driven out print.

The need for people to be able to formulate and express their thoughts to others and to understand the thoughts of others remains. All else is mechanics.

We have seen the progression of communication media go from sound and gesture, to drawing, to writing and the alphabet. Then, printing was added. Then transportation by ship, wagon, train, automobile, airship, and airplane. Before the latter two, came the electric telegraph and then the telephone. Just about coincident with the airplane came electronic communication, first as radio. Radio as a medium of transmission enabled television and wireless telephony. Both were enhanced by communication satellites. Finally, well not *finally*, we got the Internet. It gave us, in effect, another means of communication but it also holds out the possibility of complete convergence of all our current communications media. In the future we may simply turn on the communication machine then decide whom we want to communicate with and in what form—pictures, text, sound, whatever. National borders will be of little import in distance communication. Distance will be of little import in the cost.

Indeed, with the wave of change still to come, Gutenberg's contribution may yet be seen as a mere ripple. But with all this promise comes the need for responsibility. A recent publication of the Vatican's Pontifical Council for Social Communications stated:

> An individual can ascend to heights of human genius and virtue, or plunge to the depths of human degradation, while sitting alone at a keyboard and screen. Communication technology constantly achieves new breakthroughs, with enormous potential for good and ill. As interactivity increases, the distinction between communicators and recipients blurs.[7]

Let us hope we reach for the heights, not the depths.

Notes

1. Borgman, *From Gutenberg*; Brown and Duguid, *The Social Life*; Cherniak, Deegan, and Gibson, *Beyond the Book*; Cole, *Books in Our Future*; Negroponte, *Being Digital*; Toffler, *The Third Wave*.
2. Bush, "As We May Think."
3. Agre and Rotenberg, *Technology and Privacy*; Foerstel, *Freedom of Information*.
4. Negroponte, *Being Digital*.
5. Guttag, "Communications Chameleons"; Negroponte, *Being Digital*.

6. The law of diminishing returns is not an actual law, but a principle. Typically, adding more investment in effort, money, person-power, or other resource increases output or performance, but continued adding may eventually result in the opposite outcome, diminished returns, as excess resources get in each other's way. More or less, "Too many cooks spoil the broth."
7. "Ethics in Communications."

14

One Hundred Dates to Remember

Dates prior to 1600 CE are usually not known exactly, sometimes not to the century or even millennium. Dates after 1600 are occasionally disputed but rarely by more than a year or so.

BCE

50,000	Cave drawings and markings.
15,000	Humans develop speech.
8000 -3000	Writing developed in Sumer, first as three-dimensional tokens, later as marks on a flat surface.
3500 -3000	Alphabets developed in the Middle East by various Semitic groups. The Akkadian (Phoenician) version led to the Greek, Hebrew, and eventually Latin alphabets.
2000	Papyrus introduced as writing material in Egypt.

CE

100	Paper developed in China.
300	The *codex*, present form of book developed, using parchment pages.
700	Paper use spreads to Middle East.

1100 Knowledge of papermaking arrives in Europe from the Arab world.

1454 Moveable-type printing press developed in Germany by Johann Gutenberg.

1600 Word *electricity* coined by William Gilbert in England.

1753 C. M. proposes a form of electric telegraph based on using a separate line for each letter of the alphabet.

1791 Chappé brothers in France build a semaphore based on wooden arms attached to towers, the signals to be relayed from tower to tower by attendants.

1802 First steam railway locomotive built in England but not until 1825 was the concept successfully operational.

1807 First successful steamship, the *Clermont*, built by American Robert Fulton.

1819 Hans Christian Oersted, in Denmark, demonstrates the relationship of electricity and magnetism, the principle of the electromagnet.

1822 Joseph Nicéphore Niépce, in France, develops the precursor of photography.

1825 Russian Baron Schilling demonstrates an early electric telegraph that did not require a separate line for each letter as C. M. had proposed in 1753.

1837 Charles W. Cooke and William F. Wheatstone patent an electric telegraph in England.
      Samuel F. B. Morse applies for a United States patent for an electric telegraph.

1838 Cooke and Wheatstone install their telegraph for the Great Western Railway.

1839 Louis Jacques Mande Daguerre invents the daguerreotype, the beginning of modern photography.

1843 Alexander Bain, in Scotland, invents an early model of what later became the facsimile, to transmit images by telegraph.

1844 Samuel F. B. Morse demonstrates his electric telegraph to the U.S. Congress, using a line from Baltimore to Washington. Shortly after, he finally receives a U.S. patent.

1846 Royal E. House develops a printing telegraph only a few
      months after Morse's demonstration. It drew power
      from a foot treadle. (Date varies in different sources
      because of frequent model changes.)
1849 Paul Julius Reuter forms a news service in Paris, using
      both the telegraph and carrier pigeons. In 1851 he
      moves to England, establishes the Reuters Telegraph
      Company, a descendant of which still exists and is in
      the forefront of use of electronic telecommunications
      to deliver information.
1851 The New York and Mississippi Valley Telegraph Com-
      pany formed out of several smaller companies by
      Samuel Selden and Hiram Sibley.
1856 The corporate name Western Union Telegraph Company
      is established to replace New York and Mississippi
      Valley Co. It will dominate the American telegraph
      industry for a century.
1857 Hermann von Helmholtz in Germany demonstrates the
      use of sound waves to control switching on and off of
      an electromagnet, the basis of a telephone transmitter.
1860 Pony Express begins mail service on horseback between
      St. Joseph, Missouri, and Sacramento, California.
1861 Transcontinental telegraph completed with completion
      of a section from St. Joseph to Sacramento. As a
      result, the Pony Express ceases operation.
1865 Atlantic telegraph cable completed, the work sponsored
      by businessman Cyrus W. Field, with assistance from
      the United States and British navies.
1869 Transcontinental railroad line completely across the
      United States completed.
1870 Carrier pigeons used during Franco-Prussian War to
      bring microfilmed   messages into besieged Paris.
      Similar use of pigeons dates back to ancient Rome.
      Western Union Co. transmits 9,158,000 telegraph mes-
      sages.
1872 Western Electric Company founded, largely devoted to
      producing telegraph equipment. One founder is

Elisha Gray, soon to compete with Alexander Graham Bell as claimant to recognition as the inventor of the telephone. Later, AT&T bought Western Electric, making it their principal equipment supplier.

1874   Thomas Edison develops a method of quadruplex transmission on a telegraph line: two separate message exchanges, each able to send and receive at the same time.

1876   Alexander Graham Bell applies for a patent on his telephone and produces a working model.

Transcontinental railroad line across Canada completed.

1879   Thomas Edison invents the electric light which, years later, Marshall McLuhan declares to be a communications medium.

1883   Thomas Edison produces a lightbulb-like device, which demonstrates what later became known as the *Edison effect*, that an electric current entering a vacuum on one conductor could jump across a gap to another conductor, completing a circuit. As it made no immediate contribution to his work on electric lights, he ignored the phenomenon.

One million telephones are now installed in the United States.

1884   Paul Nipkow, in Germany, develops a rotating disk through which a scene can be viewed in a scanning fashion, i.e., successive left-to-right scans of one line of the scene at a time. This is later to be used in early television transmission.

1885   First automobile powered by fossil fuel, developed in Germany by Karl Benz. This claim, like that of other inventions, is disputed in several countries.

American Telephone and Telegraph Company formed from an earlier company devoted to exploiting Bell's invention.

1887   Heinrich Hertz produces electromagnetic radiation—radio waves—and demonstrates that they travel at the speed of light.

1888 George Eastman develops a means of depositing photo-sensitive chemicals on a roll of paper, enabling cameras to be much smaller and easier to use than before. He formed the Eastman Kodak Company. Later, celluloid was used to hold the photographic emulsion. A new medium of personal communication was created.

1891 Almon B. Stowger, an undertaker, develops an automatic telephone switchboard because he felt human operators were diverting calls to his competitors.

1894 Guglielmo Marconi, in Italy, transmits a radio signal first across a room, then over a distance of 2.4 km.

1895 The Lumière brothers, in France, develop a motion picture camera and a projector.

1897 Karl Braun, in Germany, creates the cathode-ray tube, later to be a critical component of radar and television systems.

1900 Marconi receives British patent number 7777 for his method of radio transmission, later to be challenged in U.S. courts.

1901 Marconi transmits a signal containing only the Morse code symbol for "S" (• • •) from England to New-foundland.

1904 Marconi wireless is installed in 68 ships; by 1907 the number is up to 124.

John A. Fleming, a former associate of Edison, in England, notes that the Edison effect, if receiving alternating current as input, produces direct current as output. DC is needed to convert the alternating current from received radio waves to direct in order to create sound. His version of the tube became known as the *Fleming diode.*

1906 American Lee de Forest modifies the Fleming diode by adding a third element in the vacuum tube which could amplify the output current. He called it the *audion.*

Canadian Reginald A. Fessenden is first to transmit voice and music via radio, thereby paving the way for broadcast radio.

1907 Boris Rosing, in Russia, applies for a patent for a television system using a mechanical scanner and a cathode ray tube. He was unable to transmit images until 1911, and these were still not really useful. His assistant is Vladimir Zworykin, later to become a giant in television development.

1910 In this period the number of telegraph messages handled
-1917   in one year in the United States reached over 100,000,000.

1912 The ship *Titanic* sinks after striking an iceberg, even though warned of the ice via wireless. Its own radio distress signals are not heard by the closest possible rescue vessel.

1914 The Teletype is developed, based on the printing telegraph, now able to provide a printing telegraph service over telephone lines, therefore direct from originator's office to destination's office. In 1930 AT&T bought the company.

1918 Edwin H. Armstrong, while in the U.S. Army, develops the superheterodyne circuit which could detect high frequency waves and enable a radio to be accurately tuned.

Regular airmail service begins, between Washington, D.C., and New York, operated by the U.S. Army.

1919 The Radio Corporation of America (RCA) is created principally by American Marconi and General Electric Companies. David Sarnoff becomes its commercial manager.

1920 First commercial, U.S. government sanctioned radio broadcast by Westinghouse Corporation's station KDKA in Pittsburgh, presents election returns.

1922 First wirephoto, a form of facsimile, introduced and used by the London *Daily Mail*.

1923 Charles Jenkins in the United States, Edouard Belin in France, and John L. Baird in England develop primitive television systems.

1924 One million households in the United States have radios.

1925 AT&T Bell Laboratories, Inc. is formed out of previous AT&T and Western Electric laboratories. It became probably the most successful of industrial research laboratories.

1926 Philo T. Farnsworth in the U.S. demonstrates his *imagedissector*, an electronic video camera.

1927 Farnsworth produces first all-electronic television systems.

Ⅰ First televised speech in the United States, by then Secretary of Commerce Herbert Hoover.

1928 First color television image transmitted by Baird in England.

1929 Vladimir Zworykin, now with RCA in the U.S., demonstrates his *kinescope*, a television receiver.

Ⅰ Television broadcasting, on a limited basis, initially video only (no sound), begun by Baird, using the British Broadcasting Corporation, then a unit of the Post Office. Sound added in 1930.

1930 David Sarnoff becomes president of RCA, to lead that company in the commercial development of radio and later television.

1933 Edwin H. Armstrong patents the circuits needed to transmit and receive frequency modulation (FM) radio.

1935 First radar developed in England, principally by Robert A. Watson-Watt. Enough were installed by 1939 to enable the British to win the Battle of Britain by enabling controllers on the ground to guide interceptor aircraft to incoming German bombers.

1936 FM radio broadcasting authorized in the United States on an experimental basis, by the FCC. In 1940 authorization is extended to commercial broadcasting. As music becomes more important in radio, FM eventually comes to dominate broadcasting.

1937 The German passenger zeppelin *Hindenburg* burns while landing at Lakehurst, New Jersey, with great loss of life. A radio reporter, Herbert Morrison of radio station WLS in Chicago, sends out the first live report of a catastrophic event over the radio.

1938 Orson Welles produces a radio play based on H.G. Wells' novel *War of the Worlds* about an invasion of Earth by Martians, causing widespread panic among listeners not accustomed to such realism on radio.

1939 First government sanctioned commercial television broadcast in the United States, by NBC, showed President Franklin D. Roosevelt opening the New York World's Fair.

1940 Edward R. Murrow broadcasts news from London during the Battle of Britain of World War II, setting a new standard for live radio news coverage.

1941 Five hundred twenty-five line television image standard, the one still used, adopted in the United States.

1945 Arthur C. Clarke publishes a paper showing that a communications satellite is possible. The satellite would receive radio signals from the ground and relay them back, reaching a wide portion of the earth's surface.
The number of telegraph messages handled in the United States reaches its peak, 236,000,000, and begins a precipitous decline.

1946 ENIAC, the first electronic computer, developed at the University of Pennsylvania by J. Presper Eckert and John Mauchly.

1948 Cable television begins, a means of rebroadcasting television signals over wire cable in a local area, avoiding many sources of interference.

1949 Claude E. Shannon and Warren Weaver publish *The Mathematical Theory of Communication*, based on earlier journal articles. This work became the basis of modern research in communication theory.

1950 The number of households in the United States with television sets jumps from 940,000 to 3,875,000.

1951 Color television broadcasting, by Columbia Broadcasting Co., begins in the United States, although few receivers then existed that were compatible with the CBS signal.

Marshall McLuhan, a professor of English at the University of Toronto, publishes *The Mechanical Bride*, beginning a new age in media criticism and understanding.

1953 *TV Guide*, a print publication about television is founded, to become one of the largest-selling magazines in the United States and indicating that new media do not usually drive old media (print) away, but change them.

1954 A U.S. Senate committee holds hearings on charges by Senator Joseph McCarthy against some officers of the Army. Marshall McLuhan later suggests that these televised hearings led to McCarthy's downfall because it was a medium he was not adept at. This was the first major impact of television on U.S. politics and also created a precedent for broadcasting congressional hearings.

1955 John R. Pierce publishes a paper proposing a communications satellite, unaware of Clarke's earlier work, but based in part on still earlier science fiction stories, one written by himself.

1956 At long last, a telephone cable stretches across the Atlantic Ocean, nearly a hundred years after the first transatlantic telegraph cable.

Ampex Corporation creates a videotape recorder for industrial use.

First major computer system based on transmissions between computers, leading, forty years later, to the Internet. (There were probably earlier examples but histories tend not to have taken notice of this form of communications as something significant.)

1957 The Soviet Union launches the first artificial earth satel-
      lite, *Sputnik*, causing great consternation in the
      United States about the U.S. position in science. Its
      communication with the earth consisted of one mes-
      sage, a repeated series of beeps.

1960 The *Echo* communications satellite is launched by a
      team of Bell Laboratories, the Jet Propulsion Labora-
      tory, and NASA. This "bird" operated by reflecting
      signals, not re-transmitting them.

1961 Time-shared computing begins at MIT, under Fernando
      Còrbato, a method of having more than one user
      simultaneously communicating with a computer, an
      essential to the later Internet.

      AT&T develops the Data-Phone, the first modem for
      connecting computers with telephones.

1962 First transatlantic television broadcast by AT&T, from
      the United States to Britain, using the Telstar satel-
      lite.

1967 One hundred million telephones are installed in the
      United States.

1969 *Sesame Street*, an educational program for small chil-
      dren, begins its still-continuing run on television,
      demonstrating the potential of TV for education.

1971 ARPANet created under sponsorship of the U.S. Defense
      Department's Advanced Research Projects Agency.
      This evolved into the Internet.

1975 SONY Corporation and partners develop the *betamax*
      system for video recording in the home, enabling
      users to record programs as they are broadcast and
      play them back later. It also begins a new industry,
      selling pre-recorded tapes.

      The Altair 8800, a personal computer in kit form, is
      offered for sale, beginning a completely unforseen
      rush for personal computers in offices, schools, and
      homes.

1978 The *video home system* or VHS method of video record-
      ing is developed by a consortium of Japanese compa-

nies led by JVC. It quickly becomes the overwhelmingly best-seller.

1979 Cellular telephone is developed in Japan, later in the Nordic countries in 1981, and in the United States in 1983.

1983 The Internet is created, initially a means of linking computer science departments at a few universities. It is actually a voluntary standard for use by computer networks to interconnect with each other, giving computer users access to other computers all over the world.

1984 A U.S. court orders AT&T to divest itself of its local operating divisions, making them separate companies, based on antitrust allegations.

1991 CNN broadcasts live television images of the bombing of Baghdad during the Gulf War—the video equivalent of the Edward R. Murrow radio broadcasts of sixty-one years earlier.

1992 The World Wide Web is created by Tim Berners-Lee at the Centre européene de recherche nucléair (CERN) in Geneva. The Web is a network of documents accessible through the Internet.

1997 Digital video (or versatile) disk (DVD), originally developed for computers, becomes a means of recording video. Now used only for prerecorded programs, it will probably replace VHS in the future.

Radios are in 98 percent of homes in the United States.

Television sets are in 97 percent of homes in the United States.

Telephones are in 93.9% of homes in the United States.

2000 Convergence of computers, telephone, and television is becoming a reality, although not yet common.

Bibliography

Abbate, Janet. *Inventing the Internet*. Cambridge: MIT Press, 1999.

Adams, Stephen B., and Orville R. Butler. *Manufacturing the Future: A History of Western Electric*. Cambridge and New York: Cambridge University Press, 1999.

"Adventures in Cybersound, the Story of the Telegraph." [cited 26 April 2000]. Available from http://www.cinemedia.com.au /SFCV-RMIT-Annex/maughton/TELEGRAPHY_LULA,html.

Aeschylus. "Agamemnon." In *The Complete Greek Tragedies*, edited by David Greene and Richmond Lattimore, 33-90. Chicago: University of Chicago Press, 1959.

Agre, Philip E., and Marc Rotenberg. *Technology and Privacy, the New Landscape*. Cambridge: MIT Press, 1997.

Aitken, Hugh G. J. *Syntony and Spark—the Origins of Radio*. New York: John Wiley & Sons, 1976.

American Heritage Dictionary of the English Language. 3rd ed. Boston: Houghton Mifflin Co., 1992.

Astle, David. "Wampum in Pre-Columbian North America." *Mohawk Nation Drummer*, May 2000.

Baird, John Logie. *Sermons, Soap and Television*. London: Royal Television Society, 1988.

Baker, W. J. *A History of the Marconi Company*. New York: Harper & Row, 1975.

Baldwin, Gordon C. *Talking Drums to Written Word*. New York: W. W. Norton, 1970.

Banks, Arthur S. "Cross-National Time Series, 1815-1973." Computer file, 1976 [cited Nov. 2000]. Available from Inter-university Consortium for Political and Social Research, Ann Arbor, Mich.

Beebe, Lucius, and Charles Clegg. *Hear the Train Blow*. New York: E.
P. Dutton, 1952.

Berg, Jerome S. *On the Short Waves, 1923-1945: Broadcast Listening in
the Pioneer Days of Radio*. Jefferson, N.C.: McFarland, 1999.

Berners-Lee, Tim, and Mark Fishetti. *Weaving the Web*. San Francisco:
Harper, 1999.

Berners-Lee, Tim, James Hendler, and Ora Lassila. "The Semantic Web."
Scientific American 284, no. 5 (May 2001): 34-45.

Bernstein, Jeremy. *Three Degrees above Zero: Bell Labs in the Information Age*. New York: C. Scribner's Sons, 1984.

Berry, W. Turner. "Printing and Related Trades." In *A History of Technology*, Vol. 5, edited by Charles Singer et al., 683-8. Oxford: Oxford
University Press, 1958.

Blum, Andre. *On the Origin of Paper*. Translated by Harry M.
Lydenberg. New York: R. R. Bowker, 1934.

Boorstin, Daniel J. *The Discoverers*. New York: Random House, 1983.

Borgman, Christine L. *From Gutenberg to the Global Information
Infrastructure*. Cambridge: MIT Press, 2000.

Bradley, Glenn. *The Story of the Pony Express*. Edited by Waddell F.
Smith, San Rafael, Calif.: Pony Express History and Art Gallery,
1964.

Brooks, John. *Telephone*. New York: Harper & Row, 1976.

Brooks, Peter W. "Aeronautics." In *A History of Technology*, Vol. 5,
edited by Charles Singer et al., 391-413. Oxford: Oxford University
Press, 1958.

_____. *Zeppelin: Rigid Airships 1893-1940*. London: Putnam Aero-
nautical, 1992.

Brown, John Seely, and Paul Duguid. *The Social Life of Information*.
Boston: Harvard Business School Press, 2000.

Brown, Robert J. *Manipulating the Ether: The Power of Broadcast Radio
in Thirties America*. Jefferson, N.C.: McFarland & Co., 1998.

Budiansky, Stephen. *The Truth about Dogs*. New York: Viking Penguin,
2000.

Burns, Paul T. "The Complete History of the Discovery of Photography." [cited 12 Oct. 2000]. Available from www.
Precinemahistory.net/900.htm.

Bush, Vannevar. "As We May Think." *Atlantic Monthly*, July 1945,
101-108.

Busnel, Rene Guy. *Acoustic Behaviour of Animals*. Amsterdam and New
York: Elsevier, 1963.

Cameron, Donald Roderick. *An Aid to National Defence: Carrier Pigeons: A National Question*. Toronto: C. Blackett Robinson, 1890.

Camp, L. Jean. *Internet Economics*. Cambridge: MIT Press, 2000.

Cannon-Brookes, Peter. *The Painted Word, British History Painting*. Woodbridge, U.K.: Boydell Press, 1991.

Carpenter, Reginald, Peter Kalla-Bishop, Kenneth Munson, and Robert Wyatt. *Powered Vehicles*. New York: Crown Publishers, Inc., 1974.

Carrington, J. F. *Talking Drums of Africa*. New York: Negro University Press, 1969.

Chandler, Alfred D., Jr. *The Railroads: The Nation's First Big Business*. New York: Harcourt Brace & World, 1965.

Chappell, Warren, and Robert Bringhurst. *A Short History of the Printed Word*. 2nd ed. Point Roberts, Wash.: Hartley & Marks, 1999.

Cherniak, Warren, Marilyn Deegan, and Andrew Gibson. *Beyond the Book: Theory, Culture, and the Politics of Cyberspace*. Oxford: Office for Humanities Communication, Oxford University Computing Services, 1996.

Chomsky, Noam. "On Cognitive Structures and Their Development: A Reply to Piaget." In *Language and Learning. The Debate between Jean Piaget and Noam Chomsky*, edited by Massimo Piattelli-Palmarini, 35-54. Cambridge: Harvard University Press, 1980.

Clapham, Michael. "Printing." In *A History of Technology*, Vol. 3, edited by Charles Singer et al., 377-416. Oxford: Oxford University Press, 1957.

Clark, Theodore H. K. *Global Electronic Commerce*. Cambridge: MIT Press, 2000.

Clark, William Philo. *Indian Sign Language*. Philadelphia: L. R. Hamersly, 1885.

Clarke, Arthur C. *How the World Was One: Beyond the Global Village*. New York: Bantam Books, 1992.

Clarke, Thomas C. "The Building of a Railway." In *The American Railway: Its Construction, Development, Management, and Appliances*, edited by Thomas C. Clarke et al. New York: Chas. Scribner's Sons, 1889.

Coe, Lewis. *The Telegraph, A History of Morse's Invention and Its Predecessors in the United States*. Jefferson, N. C.: McFarland & Co., 1993.

Cole, John Y., ed. *Books in Our Future*. Washington, D.C.: Library of Congress, 1987.

Cole, S. M. "Land Transport without Wheels. Roads and Bridges." In *A History of Technology*, Vol. 1, edited by Charles Singer, et al., 704-15. Oxford: Oxford University Press, 1957.

Corballis, Michael C. "The Gestural Origins of Language." *American Scientist* 87, no. 2 (1999): 138-145.

Costello, Elaine. *Signing: How to Speak with Your Hands.* Toronto and New York: Bantam Books, 1983.

Couey, Anna. "The Birth of Spread Spectrum. How the 'Bad Boy of Music' and 'the Most Beautiful Girl in the World' Catalyzed a Wireless Revolution—in 1941." [cited 16 Nov. 2001] Available from www.sirius.be/lamar.htm.

Crandall, Robert W. *After the Breakup: U.S. Telecommunications in a More Competitive Age.* Washington, D.C.: Brookings Institution, 1991.

Cross, Ian. "Is Music the Most Important Thing We Ever Did? Music, Development and Evolution." In *Music, Mind and Science*, edited by Suk Won Yi, 10-29. Seoul: Seoul University Press, 1999.

Crowley, David, and Paul Heyer. *Communication in History, Technology, Culture, Society.* 3rd ed. New York: Longman, 1999.

Cullinan, Gerald. *The United States Postal Service.* New York: Praeger Publishers, 1973.

Custer, George A. *My Life on the Plains or, Personal Experiences with Indians.* New York: Sheldon and Co., 1874. Reprint, Norman, Okla.: University of Oklahoma Press, 1962.

Dana, Peter H. "Global Positioning System Overview." [cited 27 Feb 2001]. Available from www. colorado.edu/geography/gcraft/notes/gps/gps_f.html.

Danelian, N. R. *AT&T. The Story of Industrial Conquest.* New York: Arno Press, 1974.

Darwin, Charles. *The Expression of the Emotions in Man and Animals.* Republished, Oxford: Oxford University Press, 1997.

Diamond, Jared. "Blueprints and Borrowed Letters." In *Guns, Germs, and Steel*, 215-238. New York: W. W. Norton, 1997.

Dickinson, H. W. "The Steam-Engine to 1830." In *A History of Technology*, Vol. 4,. edited by Charles Singer et al., 199-213. Oxford: Oxford University Press, 1958.

Digital Television. [cited 28 May 2000]. Available from www.digitaltelevision.com.

Douglas, Susan J. *Listening in: Radio and the American Imagination from Amos 'n' Andy and Edward R. Morrow to Wolfman*. New York: Times Books, 1999.

Drucker, Peter F. "Beyond the Information Revolution." *Atlantic Monthly*, October 1999, 47-57.

Dunlap, Orrin E., Jr. *Marconi, the Man and His Wireless*. New York: The Macmillan Co., 1937.

Eco, Umberto. *The Island of the Day Before*. Translated by William Weaver. New York: Viking Penguin, 1996.

_____. *Serendipities, Language and Lunacy*. New York: Columbia University Press, 1998.

Eisenstein, Elizabeth L. *The Printing Press as an Agent of Change*. Cambridge and New York: Cambridge University Press, 1979.

Ellis, C. Hamilton. "The Development of Railway Engineering" In *A History of Technology*, Vol. 5, edited by Charles Singer et al., 322-349. Oxford: Oxford University Press, 1958.

Erickson, Don V. *Armstrong's Fight for FM Broadcasting: One Man vs Big Business and Bureaucracy*. University, Ala.: University of Alabama Press, 1973.

Everett, R. R., C. A. Zracket, and H. D. Bennington. SAGE—a Data Processing System for Air Defense. In *Proceedings of the Eastern Computer Conference*, 148-155. New York: Institute of Radio Engineers, 1957.

Field, D. C. "Internal Combustion Engine." In *A History of Technology*, Vol. 5, edited by Charles Singer et al., 157-176. Oxford: Oxford University Press, 1958.

_____. "Mechanical Road Vehicles.." In *A History of Technology*, Vol. 5,. edited by Charles Singer et al., 414-437. Oxford: Oxford University Press, 1958.

Field, Henry Martyn. *The Story of the Atlantic Telegraph*. New York: Scribner's Sons, 1992.

Fischer, Claude S. *America Calling:A Social History of the Telephone to 1940*. Berkeley: University of California Press, 1992.

Fisher, David E. *A Race on the Edge of Time: Radar—the Decisive Weapon of World War II*. New York: McGraw- Hill Book Co., 1988.

Fisher, David E., and Marshall Jon Fisher. *Tube: The Invention of Television*. Washington, D.C.: Counterpoint, 1996.

Foerstel, Herbert N. *Freedom of Information: Our Right to Know: The Origins and Applications of the Freedom of Information Act*. Westport, Conn.: Greenwood Press, 1999.

Forbes, R. J. "Roads to c. 1900." In *A History of Technology,* Vol. 4, edited by Charles Singer et al., 520-457. Oxford: Oxford University Press, 1958.

"Fulfilling the Promise" (A collection of articles about the Internet). *Scientific American,* March 1997, 50-83.

Gardiner, Robin, and Dan van der Vat. *The Riddle of the Titanic.* London: Weidenfeld & Nicolson, 1995.

Garrat, G. R. M. "Telegraphy, From Early Times to the Fall of Empires." In *A History of Technology,* Vol. 4, edited by Charles Singer et al., 644-662. Oxford: Oxford University Press, 1958.

Gelb, I. J. *A Study of Writing.* Chicago: University of Chicago Press, 1952.

A General History of the Sciences. Translated by A. J. Pomerans. Vol. 2, *The Beginnings of Modern Science from 1450 to 1800,* by Rene Taton. London: Thames and Hudson, 1963

Gernsback, Hugo. *Ralph 124C41+, a Romance of the Year 2660.* Boston: The Stratford Co., 1925. Reprint, Lincoln, Nebr.: University of Nebraska Press, 2000.

Gibbs, Philip. "Is Faster than Light Travel or Communication Possible?" January 14, 1998. [cited 15 Nov 2001] Available from http://www.desy.de/ user/projects/Physics/Relativity/SpeedOfLight/ FTL.html.

Gladwell, Malcolm. "Clicks & Mortar." *The New Yorker,* December 6, 1999, 106-15.

Goddard, Peter. "The New Ad Fad." *Toronto Star,* February 12, 2000.

Gromov, Gregory R. "History of the Internet and WWW." Available from http://www.netvalley.com/intvalweb.html.

Grosvenor, Edwin S. *Alexander Graham Bell: The Life of the Man Who Invented the Telephone.* New York: Harry Abrams, 1997.

Guernsey, Lisa. "Beyond Neon: Electronic Ink." *New York Times,* June 3, 1999.

Guillemin, Amedee. "Telephone Apparatus for Rapid Transmission." In *The Electric Telegraph, An Historical Anthology,* edited by George Shiers, 61-65. New York: Arno Press, 1977.

Gunston, Bill. *Chronicle of Aviation.* London: Chronicle Communications Ltd., 1992.

Guttag, John V. "Communications Chameleons." *Scientific American* August 1999: 58-59.

Gwynn-Jones, Terry. *Farther and Faster, Aviation's Adventuring Years, 1909-1939.* Washington, D.C.: Smithsonian Institution Press, 1991.

Harte, Lawrence, Richard Levine, and Steve Prokop. *Cellular and PCS: The Big Picture.* New York: McGraw-Hill, 1997.

Hauser, Marc D. *The Evolution of Communication.* Cambridge: MIT Press, 1996.

Held, Gilbert. *Understanding Data Communication.* Chichester, U.K.: John Wiley & Sons, 1996.

Herrman, Dorothy. *Helen Keller: A Life.* New York: Alfred A. Knopf, 1998.

"High-Definition Television (HDTV)." In *Telecommunications & Multimedia Encyclopedia.* Compact disk ed. Englewood, Colo.: Jones Digital Century, n.d.

Hirsch, Robert. *Seizing the Light: A History of Photography.* Boston: McGraw-Hill Higher Education, 2000.

Historical Statistics of the United States, Colonial Times to 1970, Part 1. Washington, D.C.: U.S. Department of Commerce, 1975.

Historical Statistics of the United States, Colonial Times to 1970, Part 2. Washington, D.C.: U.S. Department of Commerce, 1975.

"History and Development of Railway Signaling." In *American Railway Signaling Principles and Practices.* Chicago: American Association of Railroads, 1953.

The History of Western Union. [cited 30 October 2000]. Available from http://members.tripod.com/morse_telegraph_club/ comphis.htm.

Hoke, Donald R. *Ingenious Yankees.* New York: Columbia University Press, 1990.

Hooke, S. H. "Recording and Writing." In *A History of Technology*, Vol. 1, edited by Charles Singer et al., 744-73. Oxford: Oxford University Press, 1954.

Horn, Andrew G. "Speech Acts and Animal Signals". In *Perspectives in Ethology*, edited by D. H. Owings and M. D. Beecher, 347-358. Vol. 12. New York: Plenum Press, 1997.

Hudson, Heather E. *Communication Satellites: their Development and Impact.* New York: Free Press, 1990.

Hughes, Thomas P. *Rescuing Prometheus.* New York: Pantheon Books, 1998.

Ifrah, Georges. *From One to Zero, A Universal History of Numbers.* New York: Viking Penguin, 1985.

Innis, Harold A. *The Bias of Communication.* Toronto: University of Toronto Press, 1991.

————. *Empire & Communications.* Victoria, B.C.: Press Porcepic Limited, 1986.

_____. "Media in Ancient Empires." In *Communication in History,* 3rd ed., edited by David Crowley and Paul Heyer, 23-30. New York: Longman, 1999.

Irwin, Manley R. *Telecommunications America: Markets without Boundaries.* Westport, Conn.: Quorum Books, 1981.

Israel, Paul. *From Machine Shop to Industrial Laboratory.* Baltimore: Johns Hopkins University Press, 1992.

Itzkoff, Donald M. *Off the Track: The Decline of the Intercity Passenger Train in the United States.* Westport, Conn.: Greenwood Press, 1985.

Johnson, George. "In Quantum Feat, an Atom in Two Places at Once." *New York Times,* February 22, 2000, Science/Health.

Jolly, W. P. *Marconi.* London: Constable & Co., 1970.

Kahin, Brian, and Hal R. Varian. *Internet Publishing and Beyond.* Cambridge: MIT Press, 2000.

Keith, Michael C. *Talking Radio: An Oral History of American Radio.* Armonk, N.Y.: M. E. Sharpe, 2000.

Kent, P. E. "The Global Maritime Distress and Safety System." In *Proceedings of the IEE Colloquium on Marine Control, Communications and Safety, London, April 3, 1989,* 7/1-4. London: Institute of Electrical Engineers, 1989.

Kern, Stephen. "Wireless World." In *Communication in History,* 3rd ed. edited by David Crowley and Paul Heyer, 105-107. New York: Longman, 1999.

Kilgour, Frederick G. *The Evolution of the Book.* New York: Oxford University Press, 1998.

King, W. James. "The Development of Electrical Technology in the 19th Century: The Telegraph." In *The Electric Telegraph, an Historical Anthology,* edited by G. Shiers, 305 ff. New York: Arno Press, 1977.

Knox, Bernard. Introduction. to *The Odyssey,* edited by Robert Fagles, 4. New York: Penguin, 1996.

Kramer, Samuel Noah. *The Sumerians, Their History, Culture, and Character.* Chicago: University of Chicago Press, 1963.

Kupfer, Peter. "Spies in the Skies. Researchers Are Developing Tiny, Airborne Devices That Can Look and Listen As They Float." [cited 20 Nov 2000]. Available from www.sfgate.com/cgi-bin/arctic.2000/11/20/MN62513. DTL&type=science.

Ladd, Jim. *Radio Waves: Life and Revolution in the FM Dial.* New York: St. Martin's Press, 1991.

Landstrom, Bjorn. *The Ship, an Illustrated History.* Garden City, N.Y.: Doubleday, 1961.

Lardner, James. *Fast Forward*. New York: W. W. Norton, 1987.

Lewin, Leonard. *Telecommunication: An Interdisciplinary Text*. Deham,. Mass.: Artech House, 1984.

Lewis, Tom. *Empire of the Air, The Men Who Made Radio*. New York: HarperPerennial, 1991.

Licklider, J. C. R. "Man-Computer Symbiosis." *IRE Transactions on Human Factors in Electronics* HFE-11 (March 1960): 4-11.

Loewenstein, Werner R. *The Touchstone of Life: Cell Communication and the Foundations of Life*. Oxford: Oxford University Press, 1999.

Logan, Robert K. *The Alphabet Effect*. New York: William Morrow & Co., 1986.

_____. *The Fifth Language*. Toronto: Stoddard Publishing Co., 1995.

Longfellow, Henry W. "The Landlord's Tale" in "Tales of a Wayside Inn" In *The Collected Works of Henry W. Longfellow*. Edinburgh: Gall & Inglis, 1865.

Lubrano, Annteresa. *The Telegraph. How Technology Innovation Caused Social Change*. New York & London: Garland Publishing, Inc., 1997.

Mackay, Angus. *A Collection of Ancient Piobaireachd or Highland Pipe Music*. East Ardsley, U.K.: EP Publishing Ltd., 1972.

Mallery, Garrick. *A Collection of Gesture Signs and Signals of the North American Indians with Some Comparisons*, 1880. Washington, D.C. Bureau of American Ethnology, Smithsonian Institution.

Marshack, Alexander. "The Art and Symbols of Ice Age Man." In *Communication in History*, 3rd ed., edited by David Crowley and Paul Heyer, 5-14. New York: Longman, 1999.

_____. *The Roots of Civilization*. 2nd ed. Mt. Kisco, N.Y.: Moyer Bell Ltd., 1991.

Marvin, Carolyn. "Early Uses of the Telephone." In *Communication in History*, 3rd ed. edited by David Crowley, and Paul Heyer, 155-162. New York: Longman, 1999.

Maugh, Thomas II. "Was Kilroy There at the Birth of the Alphabet?" *Toronto Star*, December 12, 1999.

McEwen, Neal, "The Telegraph Office." [cited 28 Jan 2001]. Available from http://fonix.metronet.com/~nmcewen/ tel_off.html.

McKnight, Lee W., and Joseph P. Bailey. *Internet Economics*. Cambridge: MIT Press, 1997.

McLuhan, Marshall. *Understanding Media, the Extensions of Man*. Cambridge: The MIT Press, 1994.

_____. "Understanding Radio." In *Communication in History*, 3rd ed. edited by David Crowley and Paul Heyer, 251-257. New York: Longman, 1999.

McLuhan, Marshall, and Robert K. Logan. "Alphabet, Mother of Invention." *Et Cetera* 34 (December 1977): 373-83.

Meadow, Charles T. *Ink into Bits: A Web of Converging Media*. Lanham, Md.: Scarecrow Press, 1998.

Meadow, Charles T., and Weijing Yuan. "Measuring the Impact of Information: Defining the Concepts." *Information Processing and Management* 37, no. 6 (1997): 697-714.

Meier, Samuel A. *The Messenger in the Ancient Semitic World*. Atlanta, Ga.: Scholars Press, 1988.

Mendelssohn. Kurt. *The Riddle of the Pyramids*. London: Thames and Hudson, 1974.

The Mishnah. Translated by Herbert Danby. Oxford: Oxford University Press, 1933.

Mitchell, B. R. *International Historical Statistics: The Americas 1750-1993*. New York: Stockton Press, 1998.

Moore, Francis. *Travels into the Inland Parts of Africa*. London: Edward Cave, 1738.

Morton, Frederic. *The Rothschilds: A Family Portrait*. New York: Atheneum, 1962.

Moschovitis, Christos J. P., Hilary Poole, Tami Schuyler, and Theresa M. Senft. *History of the Internet*. Santa Barbara, Calif.: ABC-Clio, 1999.

Munro, Neil. "The Ongoing Tug-of-War for E-Commerce Control." *Communications of the ACM* 42, no. 10 (October 1999): 17-20.

Murr, Lawrence E., and Everaldo Ferreyra Tello. "Connecting Materials Science and Music in Steel Drums." *American Scientist* 88, no. 1 (2000): 38-45.

Naughton, John. *A Brief History of the Future*. London: Weidenfeld & Nicolson, 1999.

Nickerson, Colin. "For Ships, End of the Dotted (and Dashed) Line." *Boston Globe*, January 31, 1999.

O'Driscoll, Gerard. *The Essential Guide to Digital Set-Top Boxes and Interactive Television*. Upper Saddle River, N. J.: Prentice-Hall., 1999.

O'Neill, J. E. "The Role of ARPA in the Development of the ARPANet, 1961-1972." *IEEE Annals of the History of Computing* 17, no. 4 (Winter 1995).

Oslin, George P. *The Story of Telecommunications*. Macon, Ga.: Mercer University Press, 1992.

Owen, Bruce M. *The Internet Challenge to Television*. Cambridge: Harvard University Press, 1999.

Palme, Jacob. *Electronic Mail*. Norwood. Mass.: Artech House, 1995.

"Pencil." In *Encyclopedia Americana*, 622-3. Danbury, Conn.: Grolier, 1988.

Pierce, John R. *The Beginnings of Satellite Communications*. San Francisco: The San Francisco Press, 1968.

_____. "Don't Write, Telegraph." *Astounding Science Fiction*, 1952, 82-96.

Pinker, Steven. *The Language Instinct. How the Mind Creates Language*. New York: HarperPerennial, 1995.

Pontifical Council for Social Communications Ethics in Communications. [cited 20 Nov. 2001] Available from http://www.zenit.org.. Once at that site, successively select English, Archive, Year 2000, May 30.

Pool, Ithiel de Sola . *The Social Impact of the Telephone*. Cambridge: The MIT Press, 1977.

The President's Grand Jury Testimony. *The Washington Post*, September 22, 1998.

Pyke, Thomas N., Jr. "Time-Sharing Computer Systems." In *Advances in Computers*, Vol. 8, 1-45. New York: Academic Press, 1967.

Raby, Ormond. *Radio's First Voice: The Story of Reginald Fessenden*. Toronto: Macmillan, 1970.

Rather, Dan. *Deadlines and Datelines*. New York: Morrow, 1999.

Reagan, Ronald, and Richard G. Hubler. *My Early Life, or Where's the Rest of Me?* New York: Duell, Sloan and Pearce, 1981.

Reid, James D. *The Telegraph in America*. New York: Derby Bros., 1879. Reprint, New York: Arno Press Inc., 1974.

Reid, T. R. *The Chip: How Two Americans Invented the Microchip and Launched a Revolution*. New York: Simon & Shuster, 1984.

Robinson, Andrew. "The Origins of Writing." In *Communication in History*, 3rd ed., edited by David Crowley and Paul Heyer, 36-42. New York: Longman, 1999.

Rogers, Everett M. *Diffusion of Innovation*. 4th ed. New York: Free Press, 1995.

Rosenman, Samuel I. *Working with Roosevelt*. New York: Harper, 1952.

Rowland, Wade. *Spirit of the Web: The Age of Information from Telegraph to Internet*. Toronto: Key Porter Books Limited, 1999.

Rowlands, Peter, and J. Patrick Wilson. *Oliver Lodge and the Invention of Radio*. Liverpool, U.K.: PD Publications, 1994.

Sampson, Anthony. *Empires of the Sky*. London: Hodder and Stoughton, 1984.

Schatt, Stan, and Steven Fox. *Voice/Data Telecommunications for Business*. Englewood Cliffs, N. J.: Prentice-Hall, 1990.

Schatzer, Laro. "The Speed of Light—a Limit on Principle?" [cited 1 July 2000] Available from http://monet.physik.unibas.ch/~schatzer/space-time.html.

Schmandt-Besserat, Denise. "The Earliest Precursor of Writing." *Scientific American* 238, no. 6 (1978): 50-59.

_____. "The Earliest Precursor of Writing." In *Communication in History: Technology, Culture, Society*, 3rd ed., edited by David Crowley and Paul Heyer, 15-23. New York: Longman, 1999.

_____. *How Writing Came About*. 1st abridged ed. Austin: University of Texas Press, 1996.

Sebeok, Thomas A. *Signs: An Introduction to Semiotics*. Toronto: University of Toronto Press, 1994.

Shaffner, Taliaferro P. "History of the English Telegraph." In *The Electric Telegraph, An Historical Anthology*, edited by George Shiers, 199-202. New York: Arno Press, 1977.

Shannon, Claude E., and Warren Weaver. *The Mathematical Theory of Communication*. Urbana: University of Illinois Press, 1959.

Sheridon, N. K., and M. A. Berkovitz. "The Gyricon—a Twisting Ball Display." *Proceeding of the Society for Information Display* 18, no. 3 & 4 (1977): 289-293.

Sherman, Ken. *Data Communications, a User's Guide*. 3rd ed. Englewood Cliffs, N.J.: Prentice Hall, 1990.

Sherman, William T. *Memoirs of General W. T. Sherman*. New York: Literary Classics of the United States, Inc., 1990.

Sherwood, Morgan B. *Exploration of Alaska*. Fairbanks, Alaska: University of Alaska Press, 1992.

Shiers, George. *The Electric Telegraph: An Historical Anthology*. New York: Arno Press, 1977.

Shlain, Leonard. *The Alphabet versus the Goddess*. New York: Viking, 1998.

Sidebottom, John K. *The Overland Mail*. London: The Postal History Society, 1948.

Silbergleid, Michael, and Mark J. Pescatore, eds. *The Guide to Digital Television* 3rd ed. Reprint, New York: United Entertainment Media,

2000. Also available from www.infojump.com/cat/Technology/7072-out.html.

Singer, Charles E., J. Holmyard, A. R. Hall, and Trevor I. Williams, eds. *A History of Technology*. 5 vols. Oxford: Oxford University Press, 1954-58.

Smith, Anthony. *Television: An International History*. Oxford & New York: Oxford University Press, 1995.

Smith, Curt. *The Storytellers: From Mel Allen to Bob Costas: Sixty Years of Baseball Tales from the Broadcast Booth*. New York: Macmillan USA, 1995.

Smith, G. O. "QRM Interplanetary." *Astounding Science Fiction*, 1942, 109-28.

Soloway, Elliot, and Amanda Pryor. "Next Generation in Human-Computer Interaction." *Communications of the ACM* 39, no. 4 (April 1996): 16-18.

Solymar, Laszlo. *Getting the Message: A History of Communications*. Oxford and New York: Oxford University Press, 1999.

Sommerfelt, A. "Speech and Language." In *A History of Technology*, Vol. 1, edited by Charles Singer, et al., 85-109. Oxford: Oxford University Press, 1954.

SONY. Encyclopedia article, [cited 14 Nov 2001]. Available from www.jonesencyclo. com/encyclo/update/sony.html.

Specter, Michael. "Your Mail Has Vanished." *The New Yorker*, December 6, 1999, 96-105.

"Spielberg's Lament." *New Republic* 198, no. 12 (1988): 7-8.

Spratt, H. Philip. "The Marine Steam-Engine." In *A History of Technology*, Vol. 5, edited by Charles Singer, et al., 141-156. Oxford: Oxford University Press, 1958.

Stallings, William. *Business Data Communications*. Upper Saddle River, N.J.: Prentice Hall, 2001.

Standage, Tom. *The Victorian Internet*. New York: Berkeley Publishing Group, 1999.

Stanley, Henry M. *Through the Dark Continent or, The Source of the Nile, around the Great Lakes of Equatorial Africa*, Vol. 2. London: J. B. Magrun, 1878.

Statistical Abstract of the United States. Lanham, Md.: Bernan Press, 1999.

Stern, Ellen, and Emily Gwathmey. *Once upon a Telephone, an Illustrated History*. New York: Harcourt Brace, 1994.

Stowers, A. "The Stationary Steam-Engine, 1830-1900." In *A History of Technology*, Vol. 5, edited by Charles Singer, et al., 124-140. Oxford: Oxford University Press, 1958.

Suedfeld, Peter. Isolation, Confinement, and Sensory Depredation. *Journal of the British Interplanetary Society* 21, no. 3 (1968): 222-231.

Tanaka, Shelley. *The Disaster of the Hindenberg: The Last Flight of the Greatest Airship Ever Built*. New York: Scholastic, 1993.

Taylor, A. R. "Nonverbal Communication Systems in Native North America." *Semiotica* 13, no. 4 (1975): 329-74.

Tehanetorens. *Wampum Belts*. Onchiota, N.Y.: Six Nations Indian Museum, 1972. Reprint. Ohsweken, Ont.: Iroqrafts Ltd., 1993.

"Telephone." In *Encyclopaedia Britannica*, 9th ed., 127-135. New York and London: Funk and Wagnals, 1890.

Temin, Peter, and Louis Galambos. *The Fall of the Bell System*. Cambridge and New York: Cambridge University Press, 1987.

Thompson, R. L. *Wiring a Continent: The History of the Telegraph Industry in the United States*. Princeton: Princeton University Press, 1947.

Toffler, Alvin. *The Third Wave*. New York: Morrow, 1980.

Ullman, B. L. *Ancient Writing and Its Influence*. Toronto: University of Toronto Press, 1980.

Understanding GPS: Principles and Applications, edited by Elliot D Kaplan. Boston: Artech House, 1996.

United States General Accounting Office. Information Superhighway, Issues Affecting Development. Report to Congress, GAOIRCED-94-285, Washington, D.C.: U.S. Government Printing Office, 1994.

Vanderhaeghe, Guy. *The Englishman's Boy*. Toronto: McClelland & Stewart, 1996.

Vaneechoutte, Mario, and J. R. Skoyles."The Memetic Origin of Language: Modern Humans as Musical Primates." [cited 20 Nov. 2001] Available from www.cpm.mmu.ac.uk/jom-emit/1998/vol2/vaneechoutte_m&skoyles_jr.html.

von Baeyer, Hans Christian. "Nota Bene." *The Sciences* 39, no. 1 (1999): 12-15.

Warkentin, Germaine. "In Search of 'The World of the Other.'" *Book History* 2 (1999): 1-27.

Weatherly, Frederick E. *Piano and Gown*. London: G. P. Putnam's, 1926.

Westland, J. Christopher, and Theodore H. K. Clark. *Global Electronic Commerce*. Cambridge: MIT Press, 2000.

Williams, Robert A., Jr. *Linking Arms Together*. New York: Oxford University Press, 1997.

Yoffie, David B., ed. *Competing in the Age of Digital Convergence*. Cambridge: Harvard University Business School, 1999.

Yule, George. *The Study of Language*. Cambridge: Cambridge University Press, 1985.

Zubrow, Ezra. "Archaeologist Investigates Humans' First Musical Instrument." Recording of radio interview, September 7, 2000 [cited 30 Jan 2001]. Available from http://infoculture.cbc.ca/archives/musop_09072000_flint.phtml.

Index

ABC. *See* American Broadcasting Company
Adams, Ansel, 27
Advanced Research Projects Agency, 287-88
Aeschylus, 78-79
air defense, 284-85
aircraft: Boeing, 121; Douglas DC-3, 121-22; experimental, 121; heavier than air, 120-22; lighter than air, 115-16
airmail, 19, 121
airplanes. *See* aircraft
Akkadians and alphabet, 53
Alaska, purchase of, role of telegraph in 138
alpenhorn, 41-42
alphabet, 53-56; history of, 53-54; impact of, 54-56
America's Cup races, 191-92
American Airlines, 285-86
American Broadcasting Company, 241
American Telephone & Telegraph Co., 131, 166, 237
American Sign Language, 33, 34
Ampère, Andrê Marie, 127

Ampex Corp., 249
AMR Corporation, 286
animal-powered transportation, 96-101
Antheils, George, 216-17
antitrust suit against AT&T, 168-69
Archer, Frederick, 71
Armstrong, Edwin Howard, 204, 206, 213
ARPA. *See* Advanced Research Projects Agency
ARPANet, 287-90
Arthur, 258
artificial intelligence, 313
AT&T. *See* American Telephone & Telegraph Co.
attenuation: of waves in general, 150; of electromagnetic waves, 188
Atwater Kent Co., 208
audion, 204, 205, 226
automobiles, 118-20

Babbage, Charles, 126
Baby Bells, 169, 177n13
Babylon, Israelite captivity in, 79

bagpipes, 41-44
Bain, Alexander, 128, 131, 221, 227
Baird, John Logie, 230, 234, 237, 240
bandwidth, 185-86
Bardeen, John, 217-18
Barton, Enos N., 148
baseball broadcasting, 26, 210-11
Battle of Britain, 210
baud, 128
Baudot, Emile, 128, 129; develops code, 134-35
Baudot code, 135
BBC. See British Broadcasting Corporation
Bell Patent Association, 165
Bell Telephone Laboratories, 168, 217-18, 268-69
Bell Canada, 2
Bell, Alexander Graham, 131, 145, 147-50
Bell Telephone Company, 165
Bell Labs. See Bell Telephone Laboratories
Bell, Melville, 147
Bennett, James Gordon, Jr., 137-38, 206
Benz, Karl, 117
Berners-Lee, Tim, 292
betamax, 248-50, 306
Bing Crosby Enterprises, 249
Bishop, Billy, 120
Bolt, Beranek and Newman, 287-88, 290
Boone, Daniel, 100
Boursel, Charles, 146
Brady, Matthew, 71
Braun, Karl Ferdinand, 226

Brinkley, David, 211
British Broadcasting Corporation, 225, 239
Brittain, Walter H., 217-18
broadcasting: by radio 202-12; by television 239-43; by telephone, 158
Bush, Vannevar, 306

C. M. letter, 125-26, 128, 235
cable television, 242, 246-48
Cable News Network. See CNN
call display, 173
call waiting, 173
call block telephone service, 173
call redial, 173
camels, as transportation media, 118
camera obscura, 69-70
Canadian Broadcasting Corporation, 244
canal boats, 104-5
Cannon-Brookes, Peter, 48
Capehart Co., 240
cards, punched or tabulating-machine, 283-84
Carrington, J. F., 38
Carter Electronics Corp., 168
Carterfone, 168
cathode-ray tube, 226, 231-34
CATV. See community antenna television
cave drawings, 16-17
CBS. See Columbia Broadcasting System
CD-ROM, 250
cellular telephone. See telephone, cellular
Centre europêene de recherche nuclêair (CERN), 292

change: affected by technology, 306-10; anticipated, 310-20; social, 314-19

Chappé brothers' semaphore, 89-90

chariots, Egyptian, 118

chauffeur, 111

China: and alphabet 54; and ideographic writing, 50; and printing, 64-5; paper in, 65 use of flags in, 84

Chomsky, Noam, 33

Churchill, Winston, 198

Civil War, U.S.: photography in, 71; railroads in, 113-14; telegraph in, 136

Clarke, Thomas, 96

Clarke, Arthur C., 95-6, 266-68

CNN, 241

cocooning, 317

code system for letters, 129-30. See also Morse code, Baudot code

coherer, 190, 192

Columbia Broadcasting System, 241-42, 256

Columbus, Christopher, 103

commerce, electronic: 293-96; impact of Internet on, 298

communication: animal, 14; basic questions about, 5-8; basics of, 1-30; by transportation, 95-106; principles of, 321-22; problems of, 14

communication industries, growth of, 318, 319-20

communication satellites, 263-76; first developed, 268; impact of, 274-75

community antenna television, 247

computer disks, 61; icons, 91-2

computers and writing, 68; speed of, 312

content, of a message, 7; restrictions on, 318

context dependence, 11; independence, 11

convergence of media, 313-14

Cooke, Sir William F., 127, 129

Cooke-Wheatstone telegraph, 129, 132, 146

Corballis, Michael, 33

cost of Internet use, 282-83

Coughlin, Father, 213

Crocker First National Bank, 237, 239

Crosley Corp., 208

cross-border transfer of information, 318

Cumberland Trail, 100

cuneiform writing, 50, 63-64

Custer, General George Armstrong, 81-82, 90

da Vinci, Leonardo, 115

Daguerre, Louis, 71

dagurreotype, 71

Daily Mail, London, 227

Daily Express, Dublin, 191

Daimler, Gottlieb, 118-19

Daimler-Benz, 118

Daimler-Chrysler, 117

data, definition of, 8

de Rochas, Alphonse, 116

de Forest, Lee, 204

delay in communications by satellite, 266

Department of Defense, U.S., 287-88

digital subscriber line, 283
digital video recording, 61-62, 250
display of television images, 231-34
divestiture by AT&T of subsidiaries, 169
Dowding, Air Chief Marshall C. T., 198
DSL. *See* digital subscriber line
Duchamp, Marcel, 47
DuMont Co., 240
DVD. *See* digital video recording

e-comerce. *See* electronic commerce
e-mail. *See* electronic mail.
Early Bird satellite, 270
Eastman, George, 72
Echo satellite, 269
Eco, Umberto, 22
Edison effect, 203, 205, 225
Edison, Thomas A., 203, 206; and motion picture projector, 71; and multiplex telegraphy, 134; and stock market telegraph, 138-39, 141; and telephone development, 165; and telephone patent disputes, 150; and telephone transmitter, 155-56
education, impact of Internet on, 299-300, 319
Egypt: pyramids in, 73; role in communication, 102, 110, 118
Electric Music Industries, 237
electroluminescence, 234
electromagnetic waves, 184
electromagnetic spectrum, 186, 187

electronic commerce, 293-6
electronic mail, 290-291
electronic commerce, 293-296
electronics, 179-302; definition, 179
EMI. *See* Electric Music Industries
enchanted lyre, 146
entertainment, change in, 315
Erie Canal, 105

facsimile, 128, 226-227
Faraday, Michael, 127, 184, 191, 222
Farnsworth, Philo T., 234, 236-237
FCC. *See* Federal Communications Commission
Federal Radio Commission, 236
Federal Communications Commission, 240, 241, 243
Fenton, Roger, 71
Fessenden, Reginald Aubrey, 206-207
fiber optic cable, 170, 174-175, 270-271
Field, Cyrus W., 128, 140-41
fire as signaling medium, 77, 78-80
fireside chats, 208
Fisher, David, 198
flags, 83-87; use in China, 84
Fleming, Sir John Ambrose, 204
FM. *See* frequency modulation
Fox network, 241
Franco-Prussian War: use of balloons in, 116; use of microfilm in, 71, 306; use of pigeons in, 98-99
Franklin, Benjamin, 99

frequency of waves, 150-51
frequency modulation, 204, 208, 213-14
frequency allocation, 186-87
Fulton, Robert, 111-12

Gale, Leonard D., 131
GCA. *See* ground controlled approach
General Electric Company, 207, 212, 237
geostationary earth orbit, 263
geosynchronous earth orbit, 263, 265
gestures, as signals, 89-90
Gilbert, William, 127
Gintl, Wilhelm, 134
Global Positioning System, 272-73
globalization, impact of Internet on, 300, 317
Goldmark, Pedter, 241
Gore, Senator Albert, Sr., 277
Gore, Senator Albert, Jr., 277
GPS. *See* Global Positioning System
Graf Zeppelin, 117
gray scale, 223-24
Gray, Elisha, 147-50, 165
Great Western Railway, 129
ground controlled approach, 199
Guardian, Manchester, 225
Gutenburg Bible, 65
Gutenburg, Johann, 64-66

Hammurabi Code, 73
Hardy, Thomas, 258
Harrison, John, 104
HDTV. *See* television, high definition

Helmholtz, Hermann von, 146
Henry, Joseph, 127, 131
Hero's steam engine, 100
Herodotus, 99
hertz, as measure of wave frequency, 150
Hertz, Heinrich, 184, 222
Hindenburg zeppelin: crash of, 116; report of crash on radio, 209
history, written, 16
Hoover, Secretary of Commerce Herbert, 208
House, Royal E., 134, 141
Hubbard, Gardner Greene, 165
Huntley, Chet, 211
Hush-a-Phone, 168

IBM. *See* International Business Machines Corp
Icarus, 115
iconoscope, 238
identification friend or foe, 199
ideographic writing. *See* pictorial writing
IFF. *See* identification friend or foe
image dissector, 236
IMAX, 72
impact of the Internet and WWW, 296-301
induction, electromagnetic, 184, 191
information: availability, change in, 315-16; definition of 8-9; effect of, 8; free flow of, change in, 316; nature of, 8-14; representation of, 20-25; representation of, analog and digital, 23-25; representation

of, direct and indirect, 21-22; transfer of, 5
information highway, 277-78
information infrastructure, 277
Innis, Harold, 54, 307
integrated circuits, 217-18
Intel Corp., 218
INTELSAT, 270-71
internal combustion engines, 116-22
International Business Machines Corp., 283
Internet, 277-78, 282; growth of, 289
internet service provider, 280, 283
Iridium satellite system, 271-72
ISP. See internet service provider

Jackson, Charles T., 131
Jacquard, Joseph-Marie, 126
Jenkins, Charles Francis, 234, 236, 240
Jet Propulsion Laboratory, 269
Joyce, James, 47
JVC, 249

KDKA, 207
Keller, Helen, 14
Kilby, Jack, 218
Kingstown Regatta, 191
Kleinrock, Leonard, 288
knowledge, definition of, 8-9
Knox, Bernard, 260

L-VIS. See live video insertion system
LaGuardia, Mayor Fiorello, 213
Lamarr, Hedy, 216-17
Lands' End, 295, 320

language: artificial, 11; effect on disciplines, 57; modes of, 50; natural. See natural language; spoken, 35-37; spoken, Central African, 40-41
language: of chemistry, 58; of mathematics, 56-57; of music, 56-57
latitude, measurement of, 104
law and regulation, change in, 317-19
laws, written, 73
Lenoir, Jean-Joseph, 116
letters, evolution of, 21
Leyden jar, 127
Licklider, J. C. R., 287-88
light, speed of, 77
Lindberg, Charles, 121
linkage, earth to satellite, 265
linotype, 66-67
live video insertion system, 255-57
Lodge, Sir Oliver, 184, 193
Logan, Robert, 54, 56, 307
Long, Huey, 213
long-distance telephone lines, 163-65, 166
Longfellow, Henry W., 80
longitude, measurement of, 104
low earth orbit, 263
Lucent Technologies, Inc., 148
Lumiere, Louis, 71-72
Lumiere, Auguste, 71-72
Lumiere, Antoine, 71

Maddox, Richard L., 71
mail service, 99-100; in Persia, 99
Majors, Alexander, 100
Marconi, Guglielmo, 184-85,

190-93; and Edwin H. Armstrong, 206; demonstration of wireless, 192; establishes companies, 193; patent for tuning device, 193

Marconi Company, Ltd. 193

Marconi America merger into RCA, 212

Marshack, Alexander, 17

Massachusetts Institute of Technology, 285-87

Maxwell, James Clerk, 184, 222

May, Joseph, 230

McCarthy, Senator Joseph, 211-12

MCI. *See* Microwave Communications, Inc.

McLuhan, Marshall, 19-20, 25-27, 37, 54, 91, 211-12, 307

meaning: communication of, 12-14; of a message, 7, 9-10

media: cool and hot, 19-20; definition of, 14-20; kinds of, 15-16; of communication, 16-19; of recording, 59-63

medium earth orbit, 263

medium is the message, 25-27

Mendelssohn, Kurt, 73

Merganthaler, Ottmar, 66

message, content of, 7; meaning of, 7

microfilm, 71

microwave radio, 170

microwave transmission, 214

Microwave Communications, Inc., 168

MIT. *See* Massachusetts Institute of Technology

modem, 279

Mongolfier brothers, 115-16

Montgomery Ward, 293-94, 296

Monty, Jean C., 2

Moore, Francis, 38

Morrison, Herbert, 209

Morse, Samuel F. B., 18, 128, 131-33, 137

Morse code, 131-32, 136; no longer required at sea, 142

motion pictures, 69, 256

Motorola Company, 271

Mr. Rogers' Neighborhood, 258

multiplexing, 134

Murrow, Edward R., 209-10, 211, 212

musical instruments: drums, 38-41; flint, 38; wind, 41-44

Muybridge, Eadwaeard, 71

National Broadcasting Company, 240, 241, 256

National Aeronautics and Space Administration, 269

National Television Systems Committee, 240

natural language, 11

Nature, 226

navigation systems, 272-73

NBC. *See* National Broadcasting Company

Nelson, Theodor, 292

New York *Herald*, 137-38, 191-192, 205-6

New York and Mississippi Valley Telegraph Co., 138

Newcomen, Thomas, 109

Niépce, Joseph Nicéphore, 70

Nipkow, Paul, 228-29, 234-35

Nipkow disk, 228-29, 236

Noble, Richard, 118

noise in communication system,

159
Noyce, Robert, 218
Nude Descending Staircase, 47-48
numerals, Arabic, 57-58

Oersted, Hans Christian, 127
one number telephone service, 173
orbit of a satellite, 263-64
Otto, Niklaus, 116-17

packet switching, 280-82
Page, Dr, C. G., 146
paper, 59-63; electronic, 62
paper making, 65
papyrus, 18, 59
parchment, 59
patent, definition of, 149
patent disputes, 149-50
Paul, Saint, 12
personal communications, impact of Internet on, 297-98
personal communications systems, 170-71, 216, 217
Philco Corp., 208, 237, 240
Philips Electronics N.V., 250
Phoenicians. *See* Akkadians
phosphorescence, 234
photoelectric effect, 230
photography, 69-72
pictographic writing. *See* pictorial writing
pictorial writing, 49-52
picture elements, 224-25
Pierce, John R., 266-68
pigeons as communications media, 98-99
Pike's Peak Express Company, 100

pixels. *See* picture elements
Plato's parable, 69
Pontifical Council for Social Communications, 322
Pony Express, 18, 100-101
pope, smoke signaling election of, 83
press, medium of, 15
printing press, 64-66
printing in China, 64-65
privacy in telephone systems, 174
privacy, protection of, 311
Prometheus and fire, 77
protocol for networks, Internet as, 282
pyramids, Egyptian, 73

quipu, as form of writing, 50, 52

radar, 196-202; basic concept, 200
radio, 183-220, 309; development of, 183-5 202-208; new technologies for, 213-18
radio broadcasting, 202-12; number of stations for, 209; social effects of, 212-13
radio detection and ranging, *See* radar
radio music boxes, 206
Radio Corporation of America, 195, 208, 212, 237, 238-39, and broadcasting, 240-43; formation of, 207
radiotelephone, 215-17
railroads, 113-15; American transcontinental, 115; in U.S. Civil War, 113-14, 136
railways, Roman, 113

RCA, *See* Radio Corporation of America
Reagan, President Ronald, 210-11
Reis, Johann, 146
repeaters on telephone lines, 161
research, impact of Internet on, 296-97
Reuter, Paul Julius, news service using pigeons, 98
Revere, Paul, and fire signals, 80
Richthofen, Baron von, 120
Rickenbacker, Eddie, 120
Roberts, Lawrence, 288
Ronald, Sir Francis, 129
Roosevelt, President Franklin D., 208, 212-13, 240
Rosenman, Samuel, 208
Rosing, Boris, 237-38
Rothschild, Nathan, 98
Royal Air Force, 197-99
Russell, William H., 100

SABRE system, 285-86
SAGE. *See* Semi-Automatic Ground Environment
Samaritans and false signals, 79-80
Sanders, Thomas, 165
Santa Maria, 103
Sarnoff, David, 195, 205-6, 234, 238
scanning of a graphic image, 222-23
Schilling, Baron, 129
Scots Magazine, 125
search engine, 293
Sears Roebuck, 293, 296
selenium, 230
semaphores, 87-92

Semi-Automatic Ground Environment, 284-85
Sesame Street, 258
sextant, 103-4
Shannon, Claude E., 5, 13
Sherman, General William T., 114, 136
ships: muscle-powered, 102; sailing, 102-5; trireme, 102; used in Egypt, 102
Shlain, Leonard, 55, 307
Shockley, William B., 218
Sibley, Hiram, 138
signaling: by gesture or body language, 89-90; visual, 77-93
signals: hand, 89-90; smoke, 80-83
significance: absence of message as, 44; of a message, 44
Smith, Curt, 26
Smith, Willoughby, 230
smoke signals, 80-83
social cohesion, impact of Internet on, 301
Sommering, S. T., 128
SONY Corporation, 249-50
sound: physics of, 150-53; transmission of, 33-45
sound based technology 37-44
speed of transatlantic ships, 113
Sputnik, 268
Stanley, H. M., 38
steam power for transportation, 107-15
steam engine in Egypt, 110
steamships, 111-13 *Clermont*, 111-12; *Sirius*, 111; *United States*, 111-12
Stearns, J. B., 134
Stein, Gertrude in World War I,

119
Stephenson, William, 227
Stone, John, 193
storage battery, 127
Stowger, Almon B., 163
Stromberg-Carlson Co., 208
Sumer: and domestication of ani-
mals, 97; and the wheel, 98;
and writing 17, 49, 73
superheterodyne circuit, 204, 226
Supreme Court, U.S., 193
Sutherland, Ivan, 288
Swinton, A. A., 226, 234
symbols; as constituents of a
message, 8; choice of, 20-21
Syncom satellites, 269

Talbot, William Henry Fox, 71
Talmud, 79
taxation, 318-19
Taylor, Alan R., 21; 89-90
Taylor, Robert W. 288
telecommunication: before steam
and electricity, 31-106; nature
of, 4, 18
telegraph, 125-43, background
of, 126-28; Cooke-Wheat-
stone, 129; demonstration of
to Congress by S. F. B.
Morse, 132; development of,
128-36, 308; in U.S. Civil
War, 136; Morse, components
of, 133; multiplexing on, 134;
needle, 129-30; printing, 134;
wireless, 188-96; workings,
137
telegraph cable, undersea, 128
telegraph industry, 135-36, 137-
39, 141-42
telephone, 145-77, 308; cellular,

170, 216-17, 271-72; ex-
changes, 161-65; industry,
165-69; industry, competition
in, 175-76; invention of, 147;
lines, 158-61; networks, 163-
164; new developments in,
169-76; number of in U.S.,
167; patent issues concerning;
149-50; portable, 216; receivers, 156-58; sound-powered,
153-54; switchboard, automatic, 163; switches, 161-65;
transmitters, 155-56; wireless,
179-73, 271-72; workings of,
153-55
telephone subscribers, linkage
among, 154-55
telephone systems: automation
of, 172-73; privacy in, 174
telesensing, 311
television, 221261, 309-10; and
competition from computers,
255-57; broadcasting, 211-12,
239-43; cable, 246-48; camera, 228-30, 243-44; color,
242, 244-46; computer effects
on, 255-57; developers of,
234-39; digital, 251; direct,
273-74; display, number of
lines in, 240; high definition,
251-53; impact of, 258-60;
interactive, 254-55; invention
of, 234; mobile, 245; new
developments in, 243-57; satellite, 246-47, 273-274; scan
lines, 222, 228, 231-33, 240,
243, 251
television images, display of,
231-34
television stations, number of,

242
Telstar satellite, 269-270
Tesla, Nikola, 193
Time-sharing, computer, 285-87
Times, New York, 240
Titanic, 194-96
tokens as precursors of writing,
    50
Tomlinson, Ray, 290
transistors, 217-18
transmission, digitization of,
    312; of radio signals, 188; of
    television signals, 226-34
transmission of graphic image,
    223
transmission range, radio, 188,
    189, 190
transmitters. *See* telephone trans-
    mitters and radio transmitters
transportation as communication,
    95-106
Travelocity, 286
travois, 97
Tribune, Chicago, 209
Trojan War, 78-79
Twain, Mark, 66
twisted pair of telephone lines,
    159
typewriter, 66

Ulysses, 47
Under the Greenwood Tree, 258
undersea telegraph cable, 128
undersea cable, 128, 140-41, 270

Vail, Theodore, 131, 165
Vail, Alfred, 131, 136, 137
van Musschenbroek, Pietr, 127
Vanderhaeghe, Guy, 221
VHS. *See* video home recording

video home recording, 249
video disk recording, 250, 306
video recording, 248-50
visible speech, 147
voice mail, 173
voice spectrograph, 151-53
Volta, Alessandro, 127
von Baeyer, Hans. 57

Waddell, William B., 100
wampum as form of writing, 50-
    51
War of the Worlds radio broad-
    cast, 211
Ward, Aaron Montgomery, 293-
    94
Washington, George, coat of
    arms, 84
Waterloo, Battle of, news of, 98
Watson-Watt, Robert A., 197-
    198
Watt, James, 109
wavelength, definition of, 150
Weaver, Warren, 13-14
Welles, Orson, 211
Wells, H. G., 211
Western Electric Co., 148
Western Union Telegraph Com-
    pany, 138-39, 165
Westinghouse Corporation, 207-
    8, 237, 238
Wheatstone, Sir Charles, 127,
    129, 134
Wilkins, Arnold F., 197-98
wire, disencumbering from, 313
Wireless Telegraph and Signal
    Co., 193
Wireless Telegraph Co., 193
wireless telegraphy, 188-96;
    aboard ships, 193. *See also*,

radio
wireless telephone, 170-73
wirephoto, 227
World Wide Web, 277, 291-96
World's Fair, New York, 240
WorldCom, 168
Wright, Orville, 120
Wright, Wilbur, 120
Wright brothers, first flight 120
writing, 49-63; and computers, 68; cuneiform, 50; effect of on civilization, 72-74; pictorial, 49-52; quipu, 50, 52; tools of, 63-68; wampum, 50
WUTC. *See* Western Union Telegraph Company
WWW. *See* World Wide Web

Xerox Corporation, 62

Zenith Co., 240
zeppelin airships, 116
Zeppelin, Ferdinand von, 116
Zubrow, Ezra, 33
Zworykin, Vladimir K., 234, 237-39

About the Author

Charles Meadow has had over forty-five years experience in the computer and communications fields. He was educated in mathematics at the University of Rochester and Rutgers University, then became involved with computers. Work in information retrieval led to an interest in how people use computers and how people and computers intercommunicate. He is the author of ten previous books in related fields. He received Honorable Mention in the 1975 New York Academy of Science Children's Book Awards and the 2000 Information Science Book of the Year Award from the American Society for Information Science and Technology.

Meadow has had a varied career in industry, government, and the academic world, having worked for the U.S. National Bureau of Standards, the U.S. Office of Science and Technology, the U.S. Atomic Energy Commission, General Electric Co., IBM, and Dialog Information Services. He joined Drexel University as a professor in 1974 and the University of Toronto in 1984 where he is now professor emeritus of Information Studies.

DATE DUE

MAY 1 5 2003			
AUG 2 0 2003			
FEB 0 8 2006			